Design and Applications of Self-Assembly Aggregates - From Micelles to Nanoemulsions

Edited by Juan C. Mejuto and Mihalj Poša

Published in London, United Kingdom

Design and Applications of Self-Assembly Aggregates - From Micelles to Nanoemulsions
http://dx.doi.org/10.5772/intechopen.1001674
Edited by Juan C. Mejuto and Mihalj Poša

Contributors
Abdelaziz El-hoshoudy, Ahmet Hakan Yilmaz, Asif Nawaz, Bulend Ortac, Dasari Thrimothi, Gudikandula Krishna, Muhammad Rehan Malik, Nadia Saleem, Nouf Aljabri, Omolade Ajayi, Saliha Mutlu, Sana Javaid, Sevil Savaskan Yilmaz, Shafi Ur Rehman, Vaibhav Changediya, Vera Tchakalova, Wajiha Noor

First published in London, United Kingdom, 2024 by IntechOpen
IntechOpen is the global imprint of INTECHOPEN LIMITED, registered in England and Wales, registration number: 11086078, 167-169 Great Portland Street, London, W1W 5PF, United Kingdom

British Library Cataloguing-in-Publication Data
A catalogue record for this book is available from the British Library

Additional hard and PDF copies can be obtained from orders@intechopen.com

Design and Applications of Self-Assembly Aggregates - From Micelles to Nanoemulsions
Edited by Juan C. Mejuto and Mihalj Poša
p. cm.
Print ISBN 978-0-85466-926-4
Online ISBN 978-0-85466-925-7
eBook (PDF) ISBN 978-0-85466-927-1

For EU product safety concerns:
IN TECH d.o.o., Prolaz Marije Krucifikse Kozulić 3, 51000 Rijeka, Croatia,
info@intechopen.com or visit our website at intechopen.com.

Meet the editors

Juan Carlos Mejuto is currently a professor at the Department of Physical Chemistry, University of Vigo, Spain. He graduated with a degree in chemistry in 1992 and received his doctorate in 1996, both from the University of Santiago de Compostela, Spain. He carried out his postdoctoral studies at Denis Diderot University (Paris 7), France. Between 2003 and 2009, he was Dean of the Faculty of Sciences at the University of Vigo. He is head of the Agrarian and Food Research group (AA1) at the Ourense Campus. His research interests include mechanisms of reactivity and catalysis in homogeneous and microheterogeneous media, (stability of self-assembled aggregates, supramolecular chemistry, and food chemistry.

Mihalj Poša is a Full Professor of Physical and Organic Chemistry at the University of Novi Sad, Serbia. He has supervised seven Ph.D. theses and authored or co-authored more than 100 journal publications. He is the recipient of several awards. His scientific research interests include the study of the self-association of bile acid salts (experimental and molecular modeling), thermodynamics of mixed micelles of bile acid anions with ionic and nonionic surfactants, regular solution theory of mixed micelle formation, micellar solubilization, hydrophobicity of bile acids, and molecular interactions between micellar systems and certain drugs.

Contents

Preface

Self-assembled colloidal systems are heterogeneous mixtures in which particles of one phase, generally nanometric or micrometric in size (dispersed phase), are distributed in another continuous phase. These particles spontaneously organize themselves into ordered or functional structures due to physical and chemical interactions, such as electrostatic interactions, hydrophobic/hydrophilic interactions, Van der Waals forces, and hydrogen bonds. These microheterogeneous systems are of significant interest in various scientific and technological fields because the particles assemble into ordered structures through intrinsic interactions, often without the need for external interventions. This self-assembly results in stable dispersions that can persist for long periods and form a variety of structures, including micelles, vesicles, liquid crystals, microemulsions, or nanoemulsions.

Thanks to these characteristics, self-assembled colloidal systems can be used to create highly sophisticated materials, such as scalable chemical nanoreactors or replicas of complex biological systems. As a result, they are employed in materials science, where they serve as models for producing materials with specific properties. Additionally, they are widely used in pharmacology, biotechnology, and medicine, particularly in controlled drug release systems. Their stability and functionality also make them valuable in the food and cosmetics industries, where they directly impact product performance. Furthermore, these systems have been utilized in the production of photonic materials, sensors, and optoelectronic devices.

One of the key advantages of these microheterogeneous systems is their ability to form complex and functional structures from simple components, making the study of soft matter a thriving field that sparks great interest in the scientific community. However, challenges remain in achieving precise control over the self-assembly process and extending the long-term stability of some of these aggregates.

This book is divided into two sections on micelles and mixed micelles, and micro-emulsions and nanoemulsions. It discusses research and applications of these systems in addressing various technical difficulties. Across different chapters, we explore the characteristics and uses of these self-organized systems. We hope that readers find the insights and suggestions provided by the contributing authors both interesting and helpful.

Mihalj Poša
Faculty of Medicine,
Department of Pharmacy,
University of Novi Sad,
Novi Sad, Serbia

Juan C. Mejuto
Faculty of Science,
Department of Physical Chemistry,
University of Vigo,
Ourense, Spain

Section 1

Micelles and Mixed Micelles

Chapter 1

Formation of Polymeric Micelle-Mixed Micelles: The Drug Delivery, and Radiotherapy Applications, Interaction, and Investigation by Laser Light

Bulend Ortac, Saliha Mutlu, Ahmet Hakan Yilmaz and Sevil Savaskan Yilmaz

Abstract

Surfactant micelles consist of a hydrophilic head that faces the solvent and a hydrophobic tail that faces the core. Mixed micelles, composed of several block copolymers, enhance medication administration. This text discusses the physical-chemical properties, *in vivo* and *in vitro* performance, and unique combinations of single and mixed copolymer micelles. It also investigates multifunctional mixed micelles that may respond to various stimuli for cancer treatment. Polymeric micelle-mixed micelles exhibit potential in the field of cancer radiation therapy. The purpose of encapsulating radiosensitizers in polymeric micelle-mixed micelles is to specifically target cancer cells and minimize harm to the rest of the body. Additionally, we can engineer polymeric micelle-mixed micelles to respond to specific conditions in the tumor microenvironment, like changes in pH or enzyme activity. Imaging compounds can be incorporated into mixed micelles to track the distribution and accumulation of tumor medicine during radiation therapy. To summarize, these adaptable nanocarriers can potentially enhance cancer treatment, specifically radiation therapy. However, further investigation is necessary to optimize the use of polymeric micelle-mixed micelles for irradiation. One technique that can determine the size distribution of particles, including multiple distributions and sizes of polymer micelles, whether single or mixed, is laser light scattering.

Keywords: polymeric micelle-mixed micelles, drug delivery, radiotherapy applications, laser light interaction, nanomedicine

1. Introduction

Nanomedicine has emerged as a new field of medicine in the past 10 years. It focuses on studying nanoscale systems, namely polymeric micelles (PMs), which enhance the solubility and stability of medications that are enclosed within them. PMs possess distinct properties due to their distinctive core-shell construction. These are suspensions of amphiphilic chemicals that form themselves into a core-shell structure in water. The core is hydrophobic and can hold hydrophobic medicines, while the shell is hydrophilic and acts as a stabilizer. In addition, because of their diminutive size, PMs can be assimilated by the intestinal mucosa together with the medicine, and they convey the drug through the bloodstream to the desired therapeutic site. Furthermore, PMs enhance the pharmacokinetic characteristics of the enclosed medication, exhibit a substantial capacity for carrying the medicine, and are produced by a replicable, uncomplicated, and cost-effective process [1].

Polymeric micelles can be prepared using either natural or synthetic polymers [2]. Thanks to advancements in polymerization technology, it is now possible to create copolymers with diverse topologies and specific physicochemical properties by including hydrophilic, hydrophobic, and cationic blocks [3, 4]. Therefore, by modifying the surface charge and size of polymeric micelles, as well as the release mechanism of therapeutics, it is possible to create efficient delivery systems and optimize other factors, such as cellular uptake, endosomal escape, and cellular release, which have an impact on treatment effectiveness [5]. **Figure 1** shows the polymeric micelles.

Micelles composed of conventional surfactants are less thermodynamically stable in physiological solutions. As a result of this drawback, novel drug-delivery micelles composed of block copolymeric chains of hydrophilic and amphiphilic monomer units have been developed. In contrast to conventional micelles, polymeric micelles (PM) are more stable due to their reduced CMC value [6]. In 1984, PM was initially proposed by Ringsdorf et al. PM exhibited a category of micelles composed of block copolymers comprising monomer units that are both hydrophobic and hydrophilic. Spherical micelles are produced in aqueous environments when the chain length of the hydrophilic block is increased beyond the range of the hydrophobic portion in AB-type graft copolymers [7]. Self-assembly of amphiphilic polymers occurs in aqueous environments, resulting in the formation of a fluid or solid core-shell structure. In the case of solid core formation, the particles are referred to as nanospheres; conversely, the formation of a fluid core results in PM. Containing the hydrophobic component,

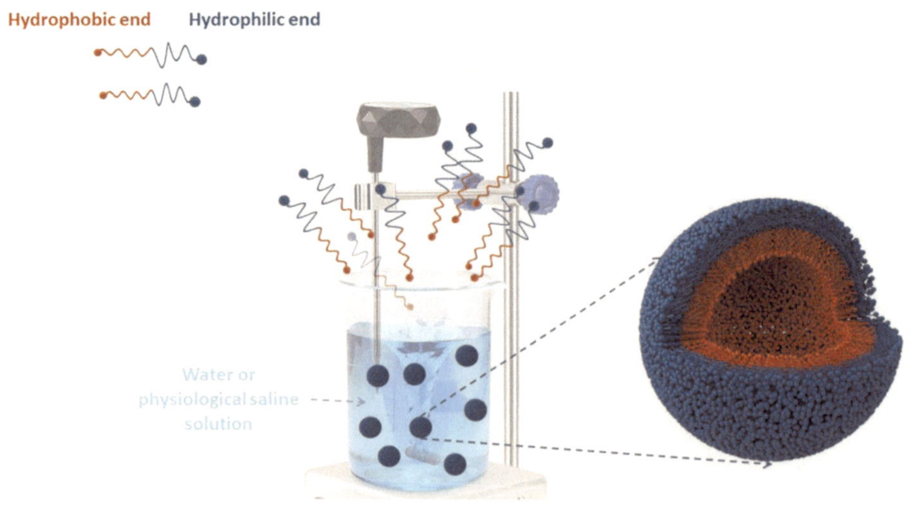

Figure 1.
Polymeric micelle structure arrangement in aqueous solution.

the nucleus of PM is a dense region. Low-water solubility pharmaceuticals are encapsulated in the core as a result of hydrophobic interactions. PM's hydrophilic outer surface is composed of an amphiphilic polymer [8]. The hydrophobic core is the primary determinant of a PM's solubility of medications that are less soluble in water. The ability of a hydrophobic core to encapsulate drugs or compounds is predominately determined by the drug molecules' compatibility with the hydrophobic internal core [9]. Core-forming components encompass a wide range of polarity and structural diversity to solubilize less soluble pharmaceuticals. The stability between pharmaceuticals and the hydrophobic core is attributed to hydrophobic interactions, which result in the formation of a thermodynamically stable complex [10]. PM's outermost layer (corona) is composed of the hydrophilic portion of the amphiphilic polymer. As for the PM, corona ought to possess effective covert property. To accommodate targeting constituents, the material must possess adequate hydrophilicity, surface charge, block copolymer chain length, and reactive group [11, 12]. Corona is accountable for safeguarding the drug against aqueous environments, facilitating its covert recognition and stability within the body *via* the reticuloendothelial system (RES), and augmenting its vascular permeability [13, 14]. PMs have enhanced biocompatibility, effective pharmacokinetics, and improved drug biodistribution as a result of these characteristics. Consequently, the drug's *in vivo* behavior can be autonomously regulated *via* the outer envelope of the inner core [15]. PM is composed of a diverse array of hydrophilic polymer blocks and typically ranges in molecular weight from 1 to 15 kDa. In general, solubilization can result in an enlargement of the hydrophobic core, which contributes to an increase in PM size. Additionally, the size of the core and corona can influence drug loading. Encapsulation efficiency can be enhanced by utilizing hydrophobic blocks that exceed the size of the core. The amphiphilic star-shaped unimolecular micelles in the PEO-b-PtBA block copolymer are capable of functioning as delivery vehicles because the hydrophobic core provides a micro-environment for the incorporation of drugs or other hydrophobic species and the hydrophilic shell stabilizes the interface between the hydrophobic core and the external medium.

Micelles are primarily colloidal dispersion sizes ranging from 5 to 100 nanometers. The formulation possesses an extremely cosmic possibility of being developed into an injectable dosage form due to its nanometric size. Cancer exhibits a vasculature that is susceptible to leakage; therefore, nanosized formulations can readily permeate the site of the tumor *via* the enhanced permeability and retention (EPR) phenomenon [16]. These substances consist of amphiphilic colloids. Using an amphiphilic or surface-active agent, these colloids are generated at a specific temperature and concentration [17]. At low temperatures and concentrations, these amphiphilic molecules coexist in an aqueous medium. However, as the concentration increases, aggregation occurs. The term for these aggregations is micelles. Critical micelle concentration (CMC) [18] refers to the concentration at which micelle formation begins; critical micellization temperature (CMT) [19] denotes the lower temperature at which amphiphilic molecules exist as unimers and no longer as aggregates. A reduced CMC value leads to the formation of stable micelles [20]. The critical micelle concentration (CMC) of mixed micelles containing sodium dodecyl sulfate (SDS) and one of five nonionic surfactants (Triton X-100, Tween 20, Tween 60, Tween 80, or Tween 85) was determined by Ćirin et al. It was discovered that nonionic surfactants whose tails are predominantly longer and more hydrophobic exhibit more robust interactions with the hydrophobic regions of SDS, thereby manifesting more robust synergism. A binary SDS–Tween 80 system exhibited the most pronounced synergistic effect. The antagonistic effect of the SDS–Tween 85 micellar system was likely due to the sterically rigid nature of the material, which is likely caused by the double bond present in its three hydrophobic tails (three C18 tails) [21]. Experimentally determined were the critical micelle concentrations of binary mixtures of surfactants and monocomponent surfactants at various molar ratios of blending. The mole fractions of building units in the binary mixed micelle, the interaction parameter, the molar surplus Gibbs energy, and the molar Gibbs energy of micellization were all computed using these values [21]. The mixed micelle structure is presented in **Figure 2**.

A spectrofluorimetric method was employed to examine the micellization of a binary mixture of surfactants Triton X100 and Brij 58 in an aqueous solution, both with and without Poloxamer 188. A probe molecule of pyrene was utilized in this investigation. An increase in temperature induces synergistic interactions among the micellar construction units in the binary mixed micelles Triton X100–Brij 58 [2]. An aqueous NaCl solution containing 0.300 mol kg^{-1} sodium chloride is utilized to examine the formation of binary mixed micelles composed of sodium cholate (SC) and hexadecyltrimethylammonium bromide (HDTAB) or sodium cholate (SC) and dodecyltrimethylammonium bromide (DTAB). The SC steroid skeleton approaches the hydrophobic interior of the micelle *via* its convex surface, as indicated by the 2D ROESY spectrum [16, 22].

In the temperature range T = (278.15–318.15) K, the micellization of a binary mixture of sodium deoxycholate (SD) and sodium cholate (SC) is investigated for an aqueous solution devoid of additives (refer to a different system for the self-association process) and for an aqueous solution containing methanol, NaCl, and methanol + NaCl. Experimentally determined were the critical micelle concentrations of binary mixtures of surfactants and monocomponent surfactants at various molar ratios of blending [23]. The observed mixed micelle exhibits greater synergy between SD-SC building units in the aqueous solution containing methanol (1.875% mol kg^{-1}) within the temperature range T = (278.15–318.15 K) compared to the mixed micelle in the aqueous solution without the additive. At elevated temperatures, building units SD-SC in a binary mixed micelle experience antagonism in an aqueous solution (without additive).

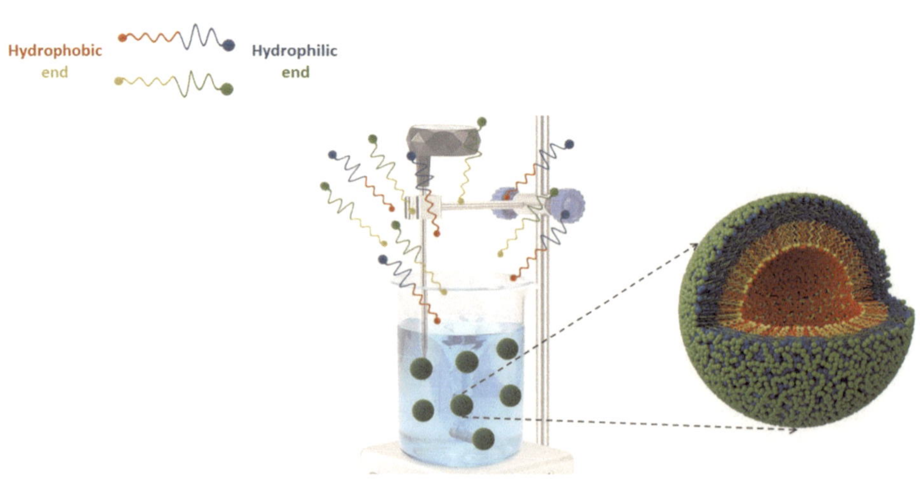

Figure 2.
Mixed micelle structure arrangement.

The process of micellization in the ternary mixture of Tween-20, sodium deoxycholate, and sodium cholate is investigated within the temperature range of T = (278.15–318.15) K. The critical micelle concentration (CMC) of ternary mixtures of the surfactants under investigation, as well as the critical micelle concentrations of binary mixtures and individual surfactants, is determined experimentally. The characteristics of the ternary mixed micelle are defined using the measured CMC values, according to the regular solution theory. The nonionic surfactant is the primary component in the ternary mixed micelle. The bile acid anions in the binary mixed micelle with Tween 20 exhibit synergistic interactions, meaning they have negative interaction coefficients of equal magnitude. Regardless of the composition of the initial combination of surfactants, sodium deoxycholate is the primary bile salt present in the mixed micelle. The inclusion of bile acid anions in the ternary micellar pseudophase is what determines which bile acid anion can decrease the overall Gibbs energy of the surfactant solution in water.

$$\Delta g^0 = RT\,(x_1 \ln cmc_1 + x_2 \ln cmc_2 + x_3 \ln cmc_3) \tag{1}$$

Eq. (1) is the molar Gibbs energy of the formation of three monocomponent micellar pseudophases that are not mixed; that is, they form separate pure micellar pseudophases. After mixing monocomponent micellar pseudophases (at constant pressure and temperature), a ternary mixed micellar pseudophase is obtained, where the change in Gibbs energy is the ideal Gibbs energy of mixing (2):

$$\Delta g_{mix}^{id} = RT(x_1\,ln\,x_1 + x_2\,ln\,x_2 + x_3\,ln\,x_3) \tag{2}$$

If a real mixed micelle contains energetic and conformational states that do not exist in monocomponent pure micelles, then there is an excess Gibbs energy compared to an ideal mixed ternary micelle (3):

$$g^e = f\left(\boldsymbol{\beta}_{ij} x_i x_j\right) \tag{3}$$

In Eq. (3) $\boldsymbol{\beta}$ is the interaction parameter from the binary mixed micelle; therefore, data are needed for all possible combinations of binary micellar pseudophases that can be derived from the ternary mixed micelle. Taking into account Eqs. (1)–(3), the change in the Gibbs energy of the formation of a real ternary mixed micellar pseudophase is:

$$\Delta g_{mM}^0 = RT\left(x_1\,ln\,cmc_1 + x_2\,ln\,cmc_2 + x_3\,ln\,cmc_3 + \Delta g_{mix}^{id} + g^e\left(\beta_{ij} x_i x_j\right)\right) \tag{4}$$

as shown in Eq. (4). The molar standard Gibbs free energy of production of the ternary mixed micelle, taking into account all three contributions, can be expressed as follows:

The nonionic surfactant dominates the composition of ternary micelles consisting of Tween 20, deoxycholic acid anion, and cholic acid anion. The molar excess Gibbs energy of micelle production is negative in every composition of the ternary mixture of surfactants. This indicates that the real ternary mixed micelles are more thermodynamically stable than the hypothetical ideal mixed micelles. The concentration of deoxycholic acid anion in the mixed micelle is significantly higher than that of cholic acid anion, by a factor of several dozen. The distribution of bile acid anions in the micellar pseudophase determines the minimization of the molar Gibbs energy of the formation of a mixed micelle. This is different from the thermodynamic stabilization of a real micelle, which is related to the excess molar Gibbs energy. The ideal ternary mixed micelle is a hypothetical concept [17].

The Gibbs-Helmholtz equation was derived from the underlying processes and describes the self-association process. The presence of the enthalpy-entropy compensation (EEC) effect requires the temperature dependence of the Gibbs free energy of micellization. The EEC parameter b, which represents the enthalpy change when entropy changes are zero, is consistently negative in the micellization process. Conversely, in the clouding point effect process, it is possible to see both positive and negative values [18].

Anions of bile acids that form primary micelles *via* the hydrophobic effect and interactions then form linear hydrophobic congeneric groups (LCG) in the lnk-lnCMC plane (where k is the RP-HPLC capacity factor and CMC is the critical micelle concentration). Bile salts that establish H-bonds in the hydrophobic domain of primary micelles deviate from the norm concerning LCG from the lnk-lnCMC plane. The formation of linear hydrophobic congeneric groups of bile acid anion derivatives is facilitated by the enthalpy-entropy compensation phenomenon and the self-association (micellization) process [19].

O/W emulsion, dialysis, solid dispersion, and the microphase separation method are a few of the numerous techniques utilized in the fabrication of polymeric micelles. A suitable organic solvent system is utilized to dissolve the polymer and substance in the solid dispersion method. A polymeric matrix was produced by evaporating the organic solvent at reduced pressure. Bag dialysis is the method employed. In summary, for the solution of polymer and drug in a water-miscible organic solvent like dimethylformamide (DMF), a reduced quantity of water is introduced while agitating continuously. The resulting mixture is subsequently dialyzed against any excess water. Emulsion formation with a continuous aqueous phase and an internal organic phase was reported using the oil-in-water emulsion solvent evaporation method. Compound emulsions are created through the simultaneous dissolution of the substance and polymer in an organic solvent that is immiscible with water, such as acetone, chloroform, tetrahydrofuran, ethanol, or a mixture thereof. Subsequently, medications are encapsulated within the inner portion of abruptly formed polymeric micelles. Utilizing organic solvents like tetrahydrofuran (THF), the microphase separation method entailed the addition of the substance and polymer. An aqueous solution was introduced dropwise while agitating both the drug and polymer solutions. To dissolve the drug and polymer, the freeze-drying process utilized an organic solvent such as t-butanol that can be freeze-dried. Following the addition of water to this solution, it is freeze-dried [24].

2. Applications for polymeric micelles (PMs)

Applications for PMs in medication delivery have advanced significantly. The main advantages of PMs are their strong core-shell structure, kinetic stability, and innate capacity to solubilize hydrophobic medications [25]. The development and remarkable advancement of polymer-based drug delivery systems and carriers have eclipsed traditional pharmaceutical methods. Intelligent drug delivery systems are being utilized, even though diffusion-controlled systems with polymer use were more common in the past. Among the most desired benefits of intelligent systems employing polymer-based drug delivery are site-specific targeting, improved bioavailability, stimuli-sensitive release, and feedback mechanisms. Numerous biocompatible polymers have been developed in response to concerns raised by biocompatibility over the widespread use of polymers in medication delivery. Additionally, novel polymers have emerged, and current polymers have been modified to fit intelligent systems thanks to chemical engineering [26–28]. There are multiple options for the production of distinct PM types, depending on the properties of the polymer and solution. PMs are formed *via* a self-assembling process that is influenced by the type of polymer employed. Different PMs can therefore be produced using di-block, tri-block, multi-block copolymers, graft polymers, stimuli-sensitive polymers, etc. Furthermore, the kind that forms can be greatly influenced by the solvent, pH, polymer concentration and ratios, cosolvent, etc. When the lipophilic portion of the amphiphilic polymer is oriented toward the micelle core and the hydrophilic portion is directed outward, the system is referred to as a PM. When reverse micelles are present, the lipophilic portion of the amphiphilic polymer is directed outward, and the hydrophilic portion is directed toward the core. Solubilizates are added to surfactant micelles to create mixed micelles. The lipid bilayer melts in the presence of surfactants, forming mixed micelles, including polar lipids and surfactants [29]. Polymeric micelles are encased in a single hydrophilic shell and have many compartments on their hydrophobic cores. Micelles can be classified into a variety of shapes according to their micellar architecture, including normal, snowman, two-hemisphere, cylindrical, reverse, dumbbell, and so forth. Additionally, multicompartment systems exhibit a variety of shapes, including disk-, worm-, raspberry-, and sheet-like forms [30]. A subclass of these nanostructures known as micelles is produced above a certain concentration (called the critical micelle concentration, or CMC) and often take the form of a sphere [31]. These structures consist of a hydrophilic corona that acts as a barrier against the aqueous environment and a hydrophobic core that is perfect for holding onto insoluble materials [32, 33]. In the case of polydisperse block copolymers (having a distribution of MW and block length), mixed micelles—that is, micelles incorporating more than one type of amphiphile—are unavoidable [34]. However, they can also be purposefully formed by mixing different block copolymers to adjust the solution properties and achieve the desired micelle structure/composition, and function [35]. For instance, in pharmaceutical applications, mixed micelles made up of the relatively hydrophilic Pluronic F127 PEO-PPO-PEO block copolymer and the relatively hydrophobic Pluronic P123 are used. The hydrophobic P123 creates an environment that is favorable for the solubilization of drug molecules that are insoluble in water, while the long PEO chains of F127 provide stability and even "stealth" qualities in aqueous solutions [36]. The ingredient lists of numerous consumer products witness the fact that low-MW surfactants are frequently used to generate mixed micelles [37, 38]. Ionic and nonionic surfactant combinations that work together to lower the critical micellization concentration (CMC) and accomplish other advantageous qualities are common examples. Block copolymers with low-MW surfactants form mixed micelles, which have a variety of possible uses and distinct mechanisms of association. Interactions between surfactants, specifically sodium dodecylsulfate (SDS), and Pluronic PEO-PPO-PEO block copolymers have been explored more thoroughly [39–42]. Kancharla and colleagues examined the structure and composition of mixed micelles that resulted from the interaction of ionic surfactant sodium dodecylsulfate (SDS) with poly(ethylene oxide)-poly(propylene oxide)-poly(ethylene oxide) (PEO-PPO-PEO) block copolymers, also known as pluronics or poluxamers, in an aqueous solution. The analysis was conducted using small-angle neutron scattering (SANS) intensity data obtained at varying contrasts. Various polymers and their concentrations have been investigated. The micelle core of the SDS + Pluronic mixed micelles is predominantly made up of alkyl chains and contains polymers and a small amount of water. Although this differs from earlier reports, it is in line with other experimental findings. The structure of SDS + Pluronic P123 (EO19PO69EO19) assemblies is being reported for the first time here. Here, interactions between the various components are examined about the impacts on the composition, surfactant concentration, and polymer hydrophobicity of the mixed micelle structure [43]. **Figure 3** shows the schematic illustration drug-loaded polymeric micelles.

Applications for amphiphilic polymer and surfactant mixtures are numerous and include drug delivery systems, detergents, cosmetics, and pharmaceuticals. Many unanswered problems surround the effects of surfactants on the structure and, in particular, the kinetics of block copolymer micelles as well as the factors that govern the solubilization kinetics. Using small-angle X-ray/neutron scattering, Myhre et al. investigated the stability and solubilization kinetics of block copolymer micelles with the addition of the surfactant sodium dodecyl sulfate (SDS). Two types of amphiphilic polymers have been used to study the surfactant's ability to dissolve polymer micelles or create mixed micelles: poly(ethylene-alt-propylene)-poly-(ethylene oxide) (PEP1-PEO20) and n-alkyl-functionalized PEO (C28-PEO5). It is known that PEP1-PEO20 micelles are frozen for a realistic timeframe, whereas the exchange kinetics of C28-PEO5 micelles are on the order of hours. In this work, we demonstrate that, even after a prolonged duration, the addition of SDS to PEP1-PEO20 results in almost little solubilization. On the other hand, Myhre et al. report that micellar disintegration and mixed micelle formation happen on an hours-long timeline after adding SDS to C28-PEO5 micelles. To get detailed structural parameters over time, the SAXS data were analyzed using a coexistence model of mixed and pristine micelles. They first notice a quick stage of fragmentation or fission, which is followed by a gradual process of reformation. At low-volume fractions, the latter process is virtually concentration-independent, but at higher concentrations, it accelerates significantly. This could

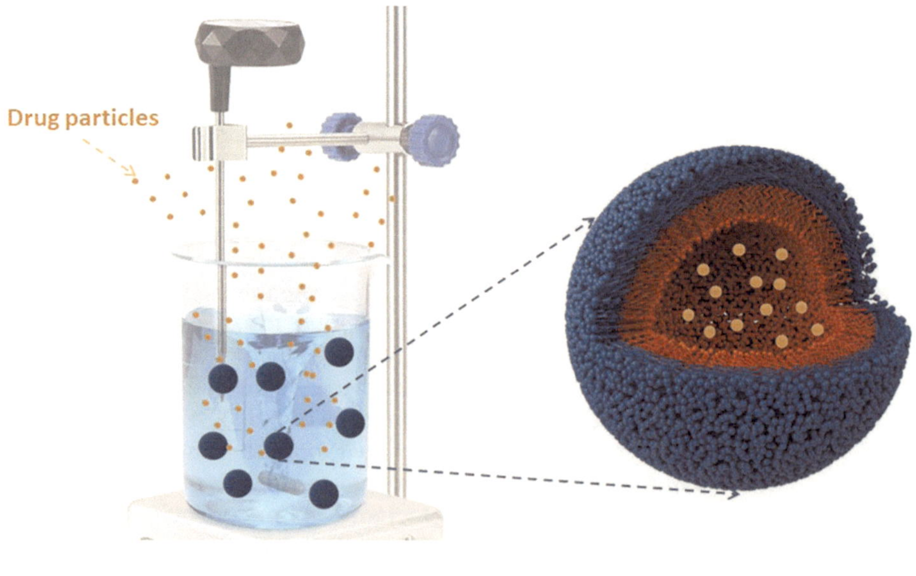

Figure 3.
Schematic illustration of the drug-loaded polymeric micelles.

signify a shift from molecular exchange mechanisms dominating the system to fusion/fission activities [44]. They have analyzed the creation of mixed micelles over time following the addition of a surfactant to the solutions of polymer micelles using a coexistence model for mixed micelles, surfactant micelles, and free surfactant chains. They discovered that solubilization with SDS did not occur for the polymer micelles or the PEP1-PEO20 system studied in this work for the SDS concentrations and period evaluated (5 days). The hydrophobic block's length and strong interfacial tension are probably to blame for this. Depending on the SDS content, mixed micelles with SDS were generated over 5–10 hours for hybrid micelles or C28–PEO5. It was discovered that the breakdown process was a combination of two processes after it was examined using a kinetic model that included two exponential functions. It was discovered that the first procedure is a quick solubilization method that is heavily concentration-dependent. This step is the dominant process above the CMC and is probably an insertion/fusion-fission stage. The second process is a rearrangement stage that is mostly mediated by unimer exchange at low SDS concentrations and by fusion/fission mechanisms at higher concentrations. Despite these findings, more research is required to determine the precise mechanism underlying the first phases of the solubilization process. TRSAXS, which needs synchrotron radiation and a stopped-flow device for quick mixing, can be used to do this [44].

Polymeric micelles can dissolve hydrophobic drugs and also possess a wide range of properties for bio-applications. These properties are achieved by modifying the typically hydrophilic outer shell of the micelles, allowing for targeted drug delivery [45] and controlled drug release [46]. Micelles formed from a single type of copolymer may have shortcomings in various ways. Due to the limited number of building blocks, these micelles are constrained in terms of their chemical and compositional diversity as well as their nanoscopic structure. A well-established yet underutilized method to tackle these problems is the combination of various polymers to create mixed micelles. It is possible to enhance the current ideal characteristics and address some of the expected limits of typical polymeric micelles with minimal effort in terms of synthetic procedures. Previous studies have consistently demonstrated the significant benefits of mixed micelles, including improved stability and enhanced effectiveness in encapsulating drugs [35, 47–49]. Copolymers in aqueous conditions form mixed micelles, which are useful for creating functional polymer nanostructures. These nanostructures have several applications in nanomedicine, such as drug delivery and bioimaging. Gerardos and her colleagues discuss the fundamental aspects of designing, structuring, and understanding the physicochemical features of mixed copolymer micelles [50]. Schematic illustration of the drug-loaded mixed micelles is presented in **Figure 4**.

Mixed surfactant solutions are commonly regarded as mixes of solvents. Mixed micellar aggregates are typically regarded as uniform blends of solvents that are spread inside a solution. However, applying typical thermodynamic models used for solvent mixtures to mixed micelles is not always straightforward. In their research, Letellier et al. examined various instances of surfactant mixes to address this issue. A significant challenge is determining the molar fraction of each surfactant in the aggregate, particularly when a charged surfactant is used in the combination. This is because the potential dissociation of the mixture's components must be taken into account both in the bulk solution and within the micelle [51]. The emergent micelles exhibited superior performance compared to their precursor copolymers in most of the experimental instances, highlighting the benefits of synergistic effects. Research has shown that mixed micelle systems offer several advantages in terms of drug entrapment effectiveness. These include increased drug loading capacity and

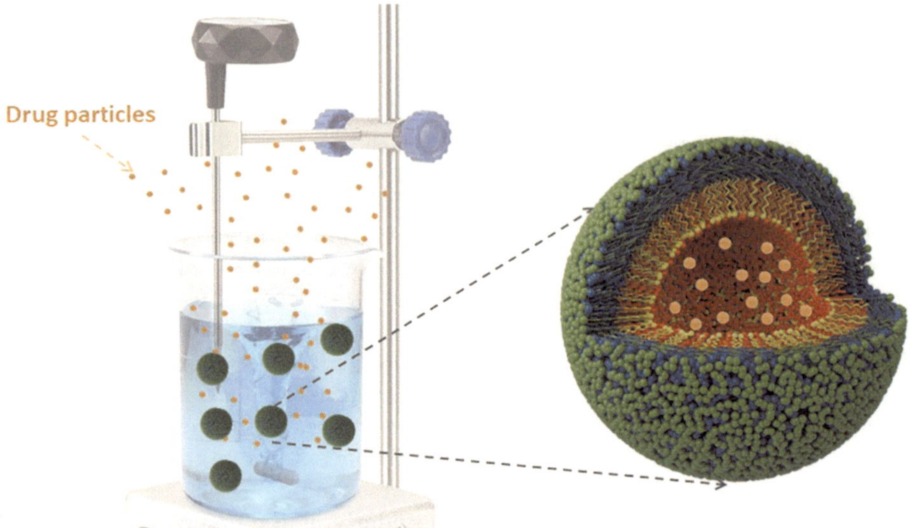

Figure 4.
Schematic illustration of the drug-loaded mixed micelles.

improved micelle durability through enhanced hydrophobic interactions, H-bonding, and other stabilizing factors. Additionally, mixed micelle systems provide better thermodynamic properties such as a decrease in critical micelle concentration and the ability to easily incorporate multiple stimuli-sensitive properties, such as pH sensitivity and temperature responsiveness. Therefore, it is imperative to conduct further clinical studies on additional nanoformulations to obtain standardized data on the behavior of vectors with similar chemical characteristics in humans. Additionally, these studies will validate their superiority over the expensive and less feasible design of new polymeric materials. In future conventional nanoformulations, polymeric mixed micelles are anticipated to incorporate a broader range of copolymers that have already been investigated in terms of their chemical composition and structure.

Research on non-viral gene carriers has made significant progress in recent decades, resulting in the development of secure and efficient vaccines utilizing lipid nanoparticles [52, 53]. Nevertheless, the modest achievements of this technique in more intricate applications emphasize the pressing need for reliable delivery technologies to properly tackle forthcoming gene delivery obstacles. Efficient delivery systems for genetic material must fulfill several criteria: (i) ensuring stable packaging; (ii) providing protection; (iii) facilitating safe carriage of the genetic payload; (iv) achieving high efficiency; and (v) maintaining a safe profile [54]. In addition to lipid-based delivery methods, cationic polymers are a promising class of materials. They have the advantage of providing a wide range of chemical options and synthetic flexibility while also having a well-defined structure and content [3, 55–57]. Amphiphilic block copolymers have become prominent in the field of biomedicine as delivery methods, primarily because of their enhanced stability and efficiency [58–60]. Amphiphilic block copolymers, which consist of a hydrophobic segment and a hydrophilic segment, can spontaneously form core-shell micelles through self-assembly [61, 62]. To facilitate the transport of genes, positively charged particles are incorporated into the outer layer to create micelleplexes, which enhance the processes of cellular uptake and endocytosis [63–65]. Nevertheless, cationic micelles can also cause lethal effects as a result of their high cationic surface charge density. By introducing a "stealth" polymer, such as poly(ethylene glycol) (PEG) or poly(ethylene oxide) (PEO), it is possible to diminish the harmful effects on cells, minimize the attachment to proteins in the blood, and prolong the duration of circulation in the bloodstream [66, 67]. Nevertheless, the advantageous impact of PEGylation is counterbalanced by a decrease in cellular absorption, a widely recognized trade-off between toxicity and effectiveness [68, 69]. Readministration of PEG antibodies in patients with a higher presence of these antibodies leads to faster blood clearance [70, 71]. Therefore, researchers are studying alternative hydrophilic polymers such as poly(N-acryloyl morpholine) (PNAM) [72, 73], poly(2-oxazoline) [74, 75], and polysarcosine [76, 77].

Lin et al. [78] developed a new type of polymeric micelles using biocompatible polymers, namely poly(ethylene glycol)-poly(lactide) (mPEG-PLA) and poly(ethylene glycol)-poly(ε-caprolactone) (mPEG-PCL). These micelles were employed as nanocarriers to encapsulate gambogenic acid (GNA). GNA-loaded mixed polymeric micelles (GNA-MMs) were produced using the cosolvent evaporation method. A solution of GNA was used to prepare mixed micelles by the cosolvent evaporation method. In summary, a mixture of mPEG2000-PLA8000 (25 mg), mPEG5000-PCL5000 copolymers (75 mg), and GNA (3 mg) was dissolved in 1.0 mL of acetone and heated to a temperature higher than the boiling point of the organic solvent. The solution was thereafter injected into the 10.0 mL of doubly distilled water. Subsequently, the concoction was vigorously agitated at ambient temperature for 2 hours, resulting in the formation of an oil-in-water emulsion. A vacuum was utilized to eliminate any remaining organic solvent and facilitate the creation of mixed micelles. After the encapsulation process, the micelle solutions were filtered using a filter [78]. Pluronic® block copolymers possess a distinctive structural characteristic that is akin to that of amphiphiles. The structure PPOmPPOnPEOm encompasses hydrophilic ethylene oxide (PEO) and hydrophobic propylene oxide (PPO) blocks, as described by Nakashima et al. A wide variety of pluronic compounds are readily accessible on the market, offering different ratios of propylene oxide (PO) to ethylene oxide (EO) and varying molecular weights. These substances include numerous unique and captivating characteristics that can be utilized in various sectors, including foaming, detergency, emulsification, dispersion, stability, and lubrication [48, 79–81]. The study conducted by Ezhilrani et al. examined the impact of a hydrophilic-hydrophilic binary micellar system, consisting of pluronic copolymers pluronic F127 and pluronic L64, on the solubilization of the antibiotic ciprofloxacin. A binary micellar system consisting of hydrophilic pluronic copolymers, specifically pluronic F127 and pluronic L64, exhibited enhanced solubilization and spontaneous binding with the antibiotic ciprofloxacin compared to the individual pluronic. The use of salt enhanced the process of drug encapsulation in the mixed micellar system [82].

Razavi et al. conducted a study on the development of a novel micelle that exhibits a responsive connection to external stimuli and possesses the requisite water solubility [83]. Using guest-host interactions, two homopolymers, polystyrene-beta-cyclodextrin (PS-β-CD) and polyethylene oxide-ferrocene (PE-FE), can form a supramolecular micelle known as PS-β-CD/PE-FE. The findings indicated that the Lennard-Jones and hydrophobic contacts exert significant influence on the process of micelle production. The study revealed that the electrical field acts as a catalyst in the reversible assembly-disassembly process of the micellar system. Additionally, we conducted a novel investigation into the interaction between PS-β-CD/PE-FE micelles and the anticancer medicines anastrozole (ANS) and mitomycin C (MIC) as a drug delivery system. The analysis of the overall energy between PS-β-CD/PE-FE micelles and pharmaceuticals indicates that the drug adsorption process is beneficial. The total energy values for the

Micelle@ANS and Micelle@MIC complexes are −638.67 and − 259.80 kJ/mol, respectively. The study's findings provide a comprehensive insight into the process of micelle synthesis, the response of micelles to electrical fields, and the behaviors of drug adsorption by micelles. This simulation investigation was conducted using classical molecular dynamics calculations. The GROMACS 5.1.4 software package is employed to conduct molecular dynamics (MD) simulations using the CHARMM force field [84]. The force field values for beta-cyclodextrin, ferrocene, polymers, and pharmaceuticals are derived from the CHARMM36 force field [85]. The solvation of each box simulation is done using TIP3P water, and any overall charge is balanced by the addition of the required number of Na^+ and Cl^- ions [86]. The temperature is regulated using the V-rescale thermostat at a constant value of 310 K, while the pressure is maintained using the Berendsen barostat at a fixed pressure of 1 bar [87–89]. The leap-frog approach is employed to solve Newton's equations of motion using a time-step of 2 femtoseconds [90]. The nonbonded contacts are limited to a radius of 1.4 nm, and the particle mesh Ewald method [91] is used to evaluate long-range electrostatic interactions. The Linear Constraint Solver algorithm [92] is used to restrict the vibrations of bonds that contain hydrogen atoms. Each simulation run has a total simulation time of 75 nanoseconds. The MD trajectory is seen and analyzed using the VMD (version 1.9.3) and GROMACS software [93]. The critical aggregation concentration (CAC) in physical units (mg/ml) can be determined using Eq. (5), allowing for a comparison with experimental data [94]:

$$\mathrm{CAC} = \frac{n}{L^3 N_A} \tag{5}$$

The variables in the equation are as follows: n represents the quantity of surfactant molecules present in the box, L is the length of one side of the simulation box, and N_A represents Avogadro's number. The PS-β-CD/PE-FE micelle is generated through the simulation of the self-assembly process. It is observed that a stable micelle maybe formed at the nanosecond scale. The total energy analysis is employed to assess the stability of the micellar system and the interactions between the host and guest polymers. Our simulations indicate that the connection of PS-β-CD/PE-FE molecules is mostly driven by the L–J interaction and hydrophobic effect. The aggregation of PS-β-CD/PE-FE is significantly influenced by coulombic and hydrogen bonding interactions. The newly developed micellar complex exhibits a high level of sensitivity to external electric fields (EF). Varying strengths of EF can effectively regulate the rate of self-disaggregation of these micelles. In the subsequent phase, the achievement of the reversible assembly-disassembly transition of the micelle was made possible by manipulating and regulating the EF. Our findings indicate that this particular EF-respond supramolecular structure has the potential to function as nano-capsules for medication delivery. The evaluation focuses on the adsorption mechanism of the PS-β-CD/PE-FE micelle and two medicines, ANS and MIC, which are placed randomly around the micelle. The drug adsorption process involves several crucial factors: Meng et al. produced a mixed micelle consisting of Pluronic F127 and D-α-tocopheryl polyethylene glycol succinate (TPGS) to enhance the transport of fluorescent dyes and proteins across the blood-brain barrier (BBB). Rhodamine 123 (Rho123) and DiR were loaded into mixed micelles made of Pluronic F127 and TPGS in a 4:1 ratio (FT). The micelles were generated by thin-film hydration. Additionally, β-galactosidase (β-Gal) was loaded into FT mixed micelles using self-assembly. The brain-targeting capacity of FT mixed micelles was assessed using both *in vitro* and *in vivo* experiments. The FT

Polymeric micelle	Drug	References	Polymeric micelle	Drug	References
mPEG-PDLLA PEG-b-PPhe	Paclitaxel	[96]	N-isopropylacrylamide (NIPAAM)	Ketorolac	[97]
MPEG-PLA	Docetaxel	[98]	Tetronic® 701 (T701) Synperonic® PE/F127 (F127), Synperonic® PE/P84 (P84)	Lornoxicam	[99]
Dextran-PLGA	Docetaxel	[100]	Poly(propylene oxide) (PPO) and poly(ethylene oxide) (PEO)	α-Tocopherol (TOC)	[101]
PLLA-mPEG	Doxorubicin	[102]	PEO–PPO–PEO	Carbamazepine	[103]
PLGA-mPEG-folate	Doxorubicin	[104]	Polyoxyethylated nonionic surfactant Pluronic® F127 (F127) cationic polyelectrolyte chitosan (CH)	Dexamethasone	[105]
PEG-PLG	Cisplatin	[106]	Egg lecithin (LE) Dexamethasone triamcinolone acetonide and	β-estradiol (E2β)	[107]
PAE-gDSMPEG)	Doxorubicin	[108]	PEG-b-PCL	Dasatinib	[109]
MPEG-PCL	Diclofenac	[110]	Poly(ethylene oxide)–poly(propylene oxide)–poly (ethylene oxide) (PEO–PPO–PEO)	Anti-apoptotic gene (bcl-XL)	[111]
Pluronic F127	Ciprofloxacin	[112]	PEO-b-PCL	dexamethasone	[113]
Poly(butylene oxide)-poly(ethylene oxide)-poly(butylene oxide)	Fluconazole	[114]	mPEG-SS-PTX/mPEG-SS-DOX	DOX/PTX	[98]
Polyhydroxyethylaspartamide (PHEA) (MePEG-b-PCL)	Dexamethasone alcohol	[115]	FA-PEG-b-(PCL-g-PEI)-b-PCL	DOX/siRNA	[116]
Methoxy(polyethylene glycol)-block-polycaprolactone	Dexamethasone acetate	[117]	FA-PCL-PHEMA-Cis-MET	PTX/MET	[118]
Methoxy poly(ethylene) glycol (MPEG)	Cyclosporine A (CsA)	[119]	mPEG poly(ethylene glycol)-block-poly(ε-caprolactone) (PEG-b-PCL)	Hypocrellin A (photosensitizer)	[120]
Poly(ethylene oxide)–poly(propylene oxide) (PEO-PPO-PEO) block copolymers	Muscone	[121]	Pluronic® F127	Zileuton™	[122]
Hexadecyloxypropyl	Cidofovir	[123]	Poloxamer	Mycophenolic acid	[124]

Table 1.
Different types of polymeric micelles used for drug delivery in the literature.

mixed micelles exhibited an average particle size of 20.03 nm and a low critical micelle concentration (CMC) of 0.0031% in water. The release of Rho123 from Rho123-loaded FT mixed micelles (FT/Rho123) in a laboratory setting demonstrated a sustained-release characteristic. FT/Rho123 exhibited superior efficacy in accumulating into brain capillary endothelial cells (BCECs) and brain tissues. The model protein β-Gal was effectively transported and stored in the brain through spontaneous loading in the FT mixed micelles. Hence, the findings suggest that F127/TPGS mixed micelles can serve as a proficient nanocarrier for delivering diagnostic and therapeutic medicines specifically to the brain [95].

There is a large amount of research available on the use of PMs for drug delivery in cancer treatment. PMs are considered a potential nanocarrier for delivering hydrophobic anticancer medicines. This information can be found in **Table 1**. Paclitaxel is derived from the bark of *Taxus brevifolia*, a tree commonly known as the Pacific yew. It exerts its effects on tubulin and hinders the process of cellular division, leading to the demise of the cell [125]. It exhibits hydrophobic properties and requires transportation *via* nanocarriers like PMs. Polymeric micelles are primarily employed to enhance the solubility of hydrophobic medicines. Therefore, polymeric micelles were utilized to improve parameters such as water solubility, efficacy, and safety profile, resulting in increased effectiveness. The toxicity of the drug administered *via* polymeric micelles was lower in comparison with the untreated paclitaxel. Kim et al. [96] documented the administration of paclitaxel medication using PMs. The author synthesized PMs using an amphiphilic diblock copolymer called mPEG-PDLLA, which stands for monomethoxy poly(ethylene glycol)-block-poly (D, L-lactide). The author then assessed the safety, pharmacokinetics, and tissue distribution of these polymeric micelles in comparison to untreated paclitaxel. Cell line investigations were conducted *in vitro* utilizing the OVCAR-3 cell line, derived from human ovarian cancer, and the MCF-7 cell line, derived from human breast cancer. The results indicated that the IC50 values of polymeric micelles in both cell lines were 0.002 μg/mL, while the IC50 values of naïve taxol were 0.002 μg/mL for ovarian cancer cell lines and 0.004 μg/mL for breast cancer cell lines. The toxicity was decreased, and the anticancer activity was enhanced [96]. **Table 1** compiles several polymeric micelles and drug categories employed for drug delivery in the literature.

3. Radiotherapy applications of polymeric micelles

3.1 Polymeric micelles for cancer therapy

Cancer poses a substantial worldwide healthcare obstacle, with surgery, chemotherapy, and radiotherapy remaining the mainstay therapeutic modalities [126, 127]. According to the World Health Organization (WHO), cancer resulted in approximately 10 million deaths worldwide in 2020, representing one out of every six deaths [128]. The most predominant forms of cancer, ranked in descending order based on the number of instances, include lung, breast, colorectal, prostate, skin, and stomach cancers. Despite significant progress in cancer research, the achievement of a cure for cancer remains a distant objective. The prevailing approach to systemic anticancer therapy involves the use of conventional mono-chemotherapy agents to treat cancer. This method involves the administration of drugs to directly eradicate cancer cells and impede their growth. However, the lack of success can be partially attributed to the use of only one method of treatment, which is unable to deal with the complex physiological conditions found in tumor microenvironments. These conditions include the existence of various types of blood vessels, impaired lymphatic drainage, increased fluid pressure between cells, and the rigid structure of the extracellular matrix. Bholakant and his colleagues [129] have shown that traditional chemotherapy remains the principal treatment for cancers. Nevertheless, it is linked to less-than-ideal outcomes and notable negative consequences. The combination of chemotherapy with numerous anticancer medications and/or other techniques has demonstrated substantial promise in addressing typical obstacles associated with conventional chemotherapy, such as tumor complexity and drug resistance. This technique enhances the efficacy of treatment, diminishes toxicity, and generates advantageous synergistic effects in the battle against cancer. Polymeric micelle-based delivery methods are currently regarded as viable choices for advanced combination approaches in cancer chemotherapy. These systems possess numerous intrinsic benefits, including a high ability to dissolve drugs, ease of modification, stable and substantial drug carrying capacity, relatively small size, high accumulation in tumors, and controlled release of drugs. A comprehensive analysis was conducted on the recent progress in multifunctional nanomaterials based on PM, which integrate several chemotherapy-based approaches such as chemocombination, chemo-gene therapy, chemo-phototherapy, and chemo-immunotherapy. The researchers [129] determined that polymeric micelle-based delivery systems show great potential as effective approaches for combining cancer chemotherapy treatments. This is due to their inherent advantages, such as their ability to solubilize drugs effectively, ease of customization, stable and high capacity for drug loading, small size, high accumulation in tumors, and controlled release of drugs. They investigated the current advancements in multifunctional nanomaterials based on polymeric micelles (PM) to combine chemotherapy-based techniques. These strategies include chemocombination, chemogene therapy, chemophototherapy, and chemo-immunotherapy. **Figure 5** shows polymeric hybrid micelle to treat tumor tissue in micel.

Polymeric micelles have emerged as a highly promising advancement in the field of chemotherapy for the treatment of cancer in the past few decades [130]. These micelles have a hydrophilic outer shell and a hydrophobic core, which gives them

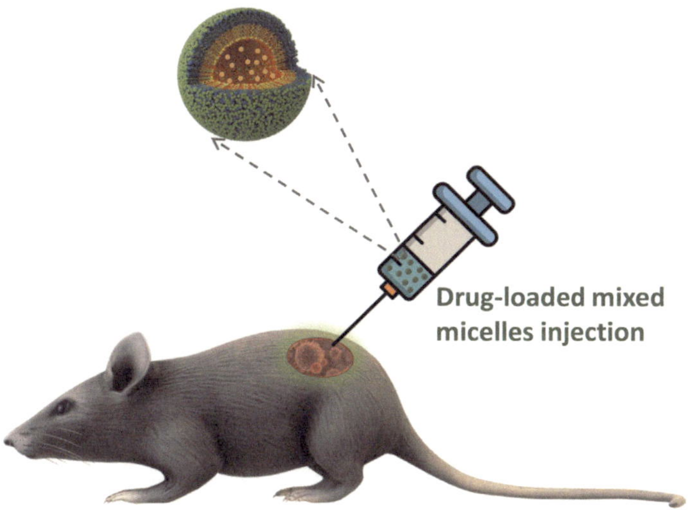

Figure 5.
Polymeric hybrid micelle to treat tumor tissue in mice.

several advantages, such as their small size, ability to self-assemble, stability in both physical and chemical aspects, compatibility with living organisms, lengthy circulation time in the bloodstream, and ability to carry and release a large number of drugs. The enhanced accessibility of building block copolymers that generate polymeric micelles also empowers researchers to fabricate the optimal drug delivery system. The production of these polymeric micelles relies on a range of non-covalent interactions, including concentration and temperature, block lengths, and preparation processes. Stimuli-sensitive micelles, a type of polymeric micelles, have gained increasing attention in recent years due to their ability to target specific sites and improve the effectiveness of medication release. Recent studies have shown that stimuli-sensitive micelles possess strong stability, high drug loading capacity, efficient drug delivery, specific site release, and minimal side effects. These micelles can adapt to the complex microenvironment found in tumors, including changes in pH, reductive responses, and enzyme activities, as well as external factors such as visible lights, UV irradiations, and temperature.

Of all the nanoparticles, polymeric micelles have recently garnered significant attention from scientists. Polymeric micelles are formed through the self-assembly of polymers and possess a distinctive amphiphilic structure. Due to the inherent physical and chemical properties of polymeric micelles, they are poised to emerge as a superior choice in the field of pharmaceuticals, particularly for the treatment of cancer. Polymeric micelles are characterized by their small size, typically measuring less than 100 nm. They will find it far more feasible to infiltrate the compact tumor and gain access to the human body. When present in human blood vessels, their tiny size will result in a more limited distribution. Therefore, they could prevent the obstruction of the circulatory system or the clumping together of particles during transportation. Chemical modifications can be applied to polymeric micelles to introduce various functional groups that serve distinct objectives.

Polymeric micelles possess an amphiphilic architecture. Because the human body is made up of mostly water, polymeric micelles can be effective carriers for hydrophobic medications. These drugs can be contained in the inner core of the micelles, where they are more soluble, while the hydrophilic outer shell of the micelles helps facilitate the entry of the drug-carrier system into the human body. To enhance the compatibility of polymeric micelles with various medications and complex solid tumors or cancer disorders, it is necessary to carry out additional chemical treatments to increase the performance of these micelles. Furthermore, achieving precise control over the conditions necessary to create an efficient micelle with diverse functional groups has emerged as a crucial objective in the field of drug administration. In addition, due to the intricate microenvironments seen in tumors and cancer sites, optimizing drug loading capacity and release selectivity has become another crucial goal in their research [130].

Majumder and his colleagues [131] discuss the design and results of different types of polymeric micelles that respond to stimuli in their study. Their goal was to address the importance of polymeric micelles in anticancer therapy. Polymeric micelles have become significant in efficiently transporting and directing hydrophobic medications. They are more readily producible in comparison with other types of nanoparticles. Although there have been notable advancements in polymeric micelles, maintaining the stability of the formulation continues to be a challenging task. The engineering of polymeric micelles has undergone tremendous advancements throughout time, resulting in improved drug loading and targeting capabilities. Several polymeric micelles have undergone therapeutic studies, and while they showed significant enhancements in animal models, their success rate in human trials has been rather low. Conversely, other formulations continue to demonstrate encouraging outcomes. In this review, we incorporated the latest advancements in the realm of polymeric micelles. However, the effective transition of a polymeric micelle from an animal model to the human patient population is still pending clinical translation. Polymeric micelles have garnered significant interest as a highly effective nanocarrier for anticancer drugs due to their ability to efficiently target drugs, selectively accumulate in tumors, exhibit biocompatibility, and offer flexible design components. This allows for manipulation of the core and shell of the micelles to suit various applications, including the use of intrinsic or extrinsic stimuli to control the release of the drug cargo. Polymeric micelles offer several advantages, such as the ability to load poorly water-soluble drugs, selective and efficient accumulation in solid tumors, potential reduction in unwanted toxicities, biodegradability and biocompatibility, and the ability to achieve controlled release of a drug over an extended period with a single administration. These benefits improve patient compliance and cost-effectiveness. Nevertheless, polymeric micelles have significant drawbacks, such as their vulnerability to degradation in a diluted environment and interaction with various biological components and proteins, as well as limited capacity for drug loading. Various techniques have been examined to address these difficulties, such as enhancing the interaction and compatibility between the micellar core and drug to achieve a high drug load and controlled release, cross-linking the micellar shell, and employing hydrophilic polymers like PEG to confer a stealth effect on the nanostructures. Several techniques exist for the synthesis of polymeric micelles. Dialysis is the prevailing technique employed at the laboratory scale to effectively eliminate organic solvents from the formulation. However, it is a time-consuming process, and scaling it up is anticipated to be challenging.

Various techniques can be used to analyze the micelles and assess their stability in a controlled environment, as well as anticipate their physical and chemical properties and stability within a living organism. Polymeric micelles have an adjustable structure that can be manipulated using a diverse range of bioactive chemicals. These molecules are specifically targeted toward the blood vessels associated with tumors and can be triggered to release drugs in response to certain stimuli. This leads to a higher level of effectiveness in the accumulation of drugs within tumors. Polymeric micelles offer a distinct advantage for delivering drugs. Nevertheless, it is crucial to meticulously evaluate the architecture and formulation of the polymer to ensure a suitable compatibility with the medicine.

The purpose of the review [132] by Afsharzadeh and her colleagues is to present a comprehensive analysis of co-delivery systems that utilize polymeric nanoparticles for cancer therapy. These systems include polymeric micelles, dendrimers, poly-d,l-lactide-co-glycolide, polyethylenimine, poly(l-lysine), and chitosan. The researchers concluded that coadministration of anticancer medications can aid in the prevention of adverse effects associated with single-agent chemotherapy, such as drug resistance in cancer. The superiority of co-delivering drugs and genes *via* nanocarriers over solitary chemotherapy for the treatment of cancer is now indisputable. Nevertheless, co-delivery approaches continue to face numerous obstacles, such as those about loading, capacity, stability, release kinetics, biocompatibility, and tumor targeting efficacy. In the present era, numerous polymeric nanoparticles (NPs) have been developed with the purpose of co-delivering anticancer pharmaceuticals. Each NP possesses distinct advantages. However, optimization of these polymeric-based nanoparticles (NPs) is necessary for their particular application, which entails controlling therapy more precisely through the targeted delivery of smaller amounts of effective anticancer agents. Furthermore, the advancement of novel copolymers as co-delivery systems provides the potential to address the shortcomings of polymeric NPs that are presently accessible. Regarding forthcoming developments in co-delivery systems, it is anticipated that comprehensive systematic inquiries will be undertaken to comprehend and optimize the circumstances governing the sequential release and efficient loading of the respective therapeutic agents. Enhancing the understanding of the factors that influence the efficacy of co-delivery strategies will soon enable logical and methodical advancements in co-delivery systems for both drugs and genes, resulting in improved therapeutic outcomes. It appears that future research should concentrate primarily on carrier-therapeutic agent interactions or interactions between therapeutic agents. When considered collectively, advancements in co-delivery systems will ultimately result in the provision of cancer patients with exceptionally efficacious therapies.

The advent of light-irradiation systems in the medical field, including endoscopic procedures, lasers, and optical fibers, has enabled the development of novel diagnostic protocols and light-sensitive drug-based selective therapies. Photochemotherapies, also known as photodynamic therapy (PDT), produce active molecular species, including singlet oxygen and free radicals, through the light irradiation of photosensitizers. These transient species are exceptionally deleterious to living organisms. Furthermore, certain photosensitizers accumulate preferentially in tissues undergoing proliferation. Utilizing this specificity property, several oncologic and ophthalmic conditions are treated.

Gibot and colleagues demonstrated in their study [133] that the physical chemistry stability of the systems under evaluation persisted regardless of dilution or aging. Due to this, their potential use as vectors in nanomedicine is critical. This stability was maintained exceptionally well in the presence of various blood components, which is consistent with their use *via* injection followed by a passive concentration increase in tumors over a few days due to their enhanced permeability and retention effect. In conclusion, the utilization of PDT on spheroids confirmed findings derived from two-dimensional cell culture and demonstrated that encapsulating Pheo in the proposed polymers significantly increased photocytotoxicity. However, minor variations in the performance of the nanovectors were emphasized: PEO(2000)-b-PCL (2800) demonstrated the highest efficiency in two-dimensional cell culture, while PEO(2400)-b-PDLLA(2000) excelled in three-dimensional experiments. One potential explanation for the discrepancy could be attributed to the varying durations of each experimental variety. As elucidated in the text, notwithstanding the favorable outcomes observed with PEO-PDLLA, its premature cargo release precludes its classification as a viable nanovector. In 2D cell culture, PEO-PCL has been demonstrated to be more effective than PEO-PS; however, in this instance as well, 3D experiments demonstrated that PEO-PS induced the greatest spheroid dissociation. Consequently, after this proof of concept, *in vivo* experiments shall be conducted to evaluate the effectiveness of said vectors for PDT in authentic physiological circumstances; this will furnish valuable insights for the *in vitro* assessments.

According to Jin et al. [134], they did a thorough review that includes many different examples of polymeric micelles being used in the field of anticancer drug delivery. However, the potential of passive targeting with polymeric micelles is constrained by the heterogeneity of the tumor microenvironment (TME). To circumvent this, polymeric micelles were endowed with a variety of functionalities. Chemical modification of the micelle-forming block copolymers permits the incorporation of diverse functionalities. As a consequence, the development of polymeric micelles capable of active targeting and triggering drug release in response to specific stimuli presented by the tumor microenvironment could be advanced. While developing functional polymeric micelles, numerous factors must be taken into account, such as systemic toxicity, targeting, drug release, therapeutic efficacy, and biodistribution. Despite rapid advancements in the laboratory, the application of polymeric micelles to clinical practice has been less than stellar.

The approval of drug delivery systems utilizing micelles is limited to a select few by the Food and Drug Administration (FDA). The prolonged time between bench and bedside translation is due to several obstacles that require meticulous investigation. The primary reason is the absence of preclinical tissue culture systems that accurately simulate biological conditions to forecast the performance of polymeric micelles in terms of therapeutic activity and targeting efficiency. Thus, additional development of evaluation instruments is required to expedite the clinical application of polymeric micelles. The advancement of multifunctional polymeric micelles has resulted in an increased complexity of their structure, which subsequently presents challenges in terms of replicable synthesis and scaling up. Clinical translation maybe impeded by the difficulty of establishing a dependable manufacturing process that meets regulatory standards. Despite the aforementioned obstacles, further research will be conducted on multifunctional micelles that are better designed, and a greater number of accumulated cases will be translated into clinical trials. Furthermore, it is imperative that forthcoming functionalities effectively capitalize on a micellar drug-free therapeutic area. By integrating chemistry, medicinal science, and clinical research, this multidisciplinary discipline has the potential to facilitate the transition of polymeric micelle-based drug delivery systems from concept to clinical practice. Nevertheless, substantial progress remains for these organizations due to the absence of comprehensive toxicity evaluations on innovative polymer micelles. As a result, newly developed micelles should be subjected to exhaustive toxicity and long-term human application testing.

For cancer-targeted drug delivery, reduction-responsive polymeric micelles were developed in a study by Wan and his colleagues [135]. These micelles were triggered with copolymers of TPGS3350-SS-DTX and TPGS3350-folate to overcome the shortcomings of conventional nanotechnology-based drug delivery systems, including short circulation time in blood, a lack of tumor targeting, and slow drug release in the interior of cancer cells, among others. The utilization of TPGS3350 as a micelle layer served to stabilize the micelles and prolong their circulation time in the blood. Tumor cells can consume these micelles *via* the folate-receptor-mediated endocytosis pathway, following their accumulation at the tumor site *via* the EPR effect. Following that, rapid drug release was accomplished within cancer cells by utilizing a high concentration of glutathione (GSH) present within the cells. The functions exhibited by the micelles developed in this research comprised active targeting, prolonged blood circulation, rapid intracellular drug release, and more. The collaborative functions observed in micelles have the potential to substantially enhance the efficacy of chemotherapeutic agents while concurrently mitigating their deleterious side effects, as evidenced by the research findings. Moreover, this work develops a drug delivery strategy to investigate the clinical viability of utilizing reduction-responsive micelles in chemotherapy. In summary, this study developed reduction-responsive polymeric micelles to achieve targeted and efficient drug delivery. Micelles containing folate were used to specifically target cancer cells. Following the uptake of TSD/TF micelles by tumor cells, the disulfide linkages in TPGS3350-SS-DTX could be cleaved by the cancer cells' abundant GSH, thereby facilitating rapid drug release. In contrast to conventional drug formulations, the anticancer activity of TSD/TF micelles was manifestly enhanced, according to the results of this study. As a consequence, clinical applications of the reduction-responsive polymeric micelles developed in their research appear promising.

Therapeutic and imaging applications have been sparked by gold nanoparticles (AuNPs). Due to their nontoxicity and nearly threefold greater X-ray attenuation per unit weight compared to iodine, AuNPs are appealing for imaging applications. AuNPs and other therapeutic agents can sensitize tumor cells to ionizing radiation. Gold-loaded polymeric micelles (GPMs) were synthesized to fabricate a nanoplatform capable of generating computed tomography (CT) image contrast, sustaining substantial tumor accumulation, and functioning as a radiosensitizer all at once. The amphiphilic diblock copolymer poly(ethylene glycol)-b-poly(ε-caprolactone) was utilized to encapsulate AuNPs measuring 1.9 nm in diameter within the hydrophobic interior of micelles. Mean hydrodynamic diameters of 25–150 nm and modest polydispersity were characteristics of the end products. GPMs enhanced the demarcation of tumor margins *via* computed tomography for 24 hours after being administered intravenously as pool contrast agents. By utilizing a radiation research infrastructure for small animals, radiation therapy was thus guided by GPM-enhanced CT imaging. The median survival time of tumor-bearing rodents was observed to be 1.7 times longer when gold was utilized in conjunction with radiation therapy, as opposed to when gold was used in isolation. It is anticipated that the application of these capabilities to support and improve the effectiveness of radiation therapy in human cancer patients could be possible. Al Zaki and colleagues [136] present the creation of a multifunctional micelle that accomplishes significant tumor accumulation, generates CT image contrast, and displays extended circulation times in animal and cellular models at sublethal radiation doses. Additionally, the micelle functions as a sensitizer for radiation therapy. Gold-loaded polymeric micelles (GPMs) with adjustable hydrodynamic dimensions between 25 and 150 nanometers were successfully synthesized *via* microemulsion synthesis. The highly concentrated clusters of 1.9 nm AuNPs are integrated into the hydrophobic core of the GPMs, which are produced by employing the amphiphilic diblock copolymer poly-(ethylene glycol)-b-poly(ε-caprolactone) (PEG-b-PCL). This is illustrated in **Figure 6**. They initially assessed the capacity of

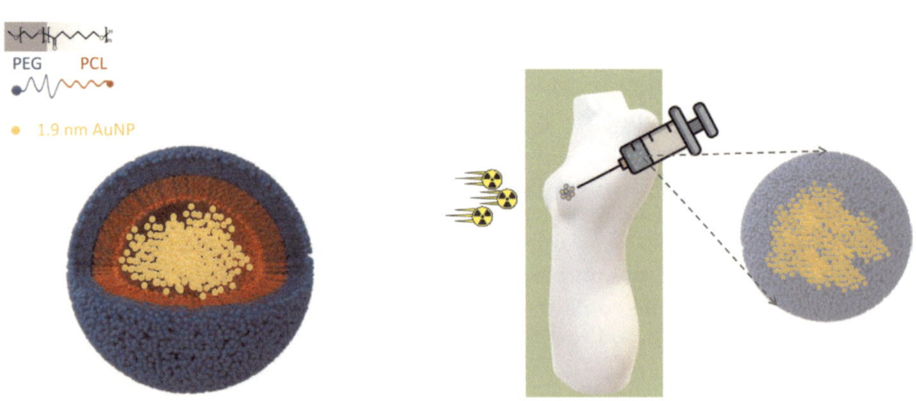

Figure 6.
Golden-loaded polymeric micelles (GPMs) are illustrated schematically. PEG-b-PCL, an amphiphilic diblock copolymer, stabilizes gold nanoparticles as they self-assemble into the hydrophobic interior of micelles. The number of individual gold nanoparticles in each GPM varies from hundreds to thousands, contingent upon their dimensions. Adopted from Ref. [136].

GPMs to accelerate radiation-induced double-stranded DNA breaks *in vitro*. The capability of GPMs to generate contrast for CT blood pools and tumor imaging was then evaluated. In conclusion, we examined whether, in murine tumor xenograft models, the radiosensitization of cells increased viability.

An agent that increases the susceptibility of cancer cells to radiation therapy is referred to as a radiosensitizer. Through the disruption of cancer cell DNA and subsequent death, radiation therapy reduces tumor size by employing high-energy radiation. Therapeutic agents that destroy or impede the proliferation of cancer cells, such as chemotherapy, or agents that exclusively induce cell death upon interaction with radiation, can function as radiosensitizers. Paclitaxel (PTX), 17-allylamino-17-demethoxygeldanamycin (17-AAG), and rapamycin (RAP) are all contained within the multidrug-loaded polymeric micelle Triolimus. In the present investigation, Tomoda and his colleagues [137] assessed the *in vitro* and *in vivo* radiosensitizing properties of Triolimus. Triolimus' radiosensitizing effects on A549 cells were dose-dependent; even at modest radiation doses (2 Gy), Triolimus inactivated significantly at 2 nm. A greater degree of radiosensitization was noted in comparison with PTX or radiation alone, with no increase in acute toxicity. Based on the findings presented, it is advisable to conduct additional research on the combination of Triolimus and radiation therapy. High drug dosage ratio Triolimus can be produced *via* a straightforward freeze-drying process. *In vitro* and *in vivo* experiments demonstrated the potential of Triolimus as an innovative radiosensitizer. While additional research is necessary to elucidate the mechanisms underlying radiosensitization, chemotherapy regimens that combine Triolimus and radiation therapy may be effective in the treatment of cancer.

Hydrogels, produced from a network of percolated micelles, can achieve drug delivery. An instance of a polymer-based system is poly(D, L-lactic acid-co-glycolic acid)-b-poly(ethylene glycol)-b-poly(D, L-lactic acid-co-glycolic acid). This polymer forms a hydrogel that is responsive to changes in temperature. The researchers found that this technique enhances the therapeutic application of gemcitabine in cancer chemoradiotherapy. Gemcitabine possesses anticancer and radiosensitization characteristics. The study reported the utilization of a gemcitabine palmityl derivative. A noteworthy characteristic is the occurrence of micelle formation in the hydrogel by this palmityl derivative of gemcitabine, which can be categorized as a polymer prodrug. These micelles and aggregates demonstrated sustained release. In addition, during *in vitro* testing, the micelles were rapidly taken up by the 4 T1 tumor cells. Through *in vivo* research, it was discovered that a solitary administration of the hydrogel-containing micelles, combined with three times the amount of X-ray exposure, effectively suppressed the tumor in the mouse model with generated tumors [138].

4. Interaction and investigation of polymeric micelles by laser light

The laser light scattering method was used to examine the thermodynamic coexistence of simple and mixed micelles in different solutions containing taurocholate (TC) and lecithin (L). The light scattering setup comprised a Lexel Co. model 95 argon-ion laser from Palo Alto, CA; a Brookhaven goniometer from Brookhaven Instruments Corporation in Holtsville, NY, equipped with a temperature-controlled cuvette holder for light scattering; and a toluene index matching bath. The experiment involved doing light scattering measurements at a wavelength of 5145 Å and a sample temperature of 37°C. Light scattering is a technique used to analyze the variations in the intensity of scattered light. It provides valuable information about the sizes and heterogeneity of the particle populations. The scattered intensity of each TC-L solution was measured and standardized by the scattered intensity from toluene when the total lipid concentration was below 10 g/dL. This significant discovery not only offers an improved method for determining the molecular mass of micelles but also provides more insights into the interpretation of micellar lipid interactions. Since conjugated cholate is the sole physiological bile salt containing three hydroxyl groups, this finding serves as a foundation for further exploration of the interactions among various simple bile salt micelles, mixed bile salt-L micelles, and coexisting simple micelles and mixed micelles [139]. Mixed micelles are formed by mixing a solution of poly(methyl methacrylate)-b-poly(ethylene oxide) (PMMAb-PEO) in toluene. This is caused by the hydrogen-bonding complexation between poly(methyl methacrylate) (PAA) and poly(ethylene oxide) (PEO). They studied the changes in mixed micelles by employing a mixture of static and dynamic laser light scattering (LLS) techniques. The LLS measurements were performed using an ALV/DLS/SLS-5022F spectrometer equipped with a multi-τ digital time correlation (ALV5000) and a cylindrical 22 mW UNIPHASE He-Ne laser (λ0 = 632 nm) as the light source. In the context of LLS, they successfully determined the weight-average molar mass (M_w), the root-mean-square radius of gyration $\langle Rg^2\rangle z^{1/2}$ (also denoted as $\langle R_g\rangle$), and the second virial coefficient A2 by analyzing the angular dependence of the absolute excess time-average scattering intensity, referred to as the Rayleigh ratio Rvv(q) [23]. Using laser light scattering, they investigated the development of PS-b-PAA/PMMA-b-PEO mixed micelles. Xie et al. investigated the relationship between the molar ratio of PEO to PAA. The analysis of their findings indicates that the interaction between PAA and PEO in the central region, as well as the separation between PS and PMMA in the outer layer, influences the development of the system. When segregation becomes the dominant factor in complexation, mixed micelle undergoes a long-term standing process and transforms into a hyperbranched structure [140].

Laser light scattering and isothermal titration calorimetric techniques were used to analyze the aqueous solution of PEO with varying amounts of SDS. Dynamic light scattering measurements were conducted using a Brookhaven BI200 goniometer and BI9000 digital correlator. The light source is an argon-ion laser with a wavelength of 488 nm that can be adjusted in power. The time correlation function was analyzed using the inverse Laplace transform of REPES, which was provided by the GENDIST software package. The addition of monovalent salt does not modify the hydrodynamic characteristics of polyethylene oxide (PEO) in an aqueous solution. The presence of a monovalent anionic surfactant, such as SDS, causes surfactant monomers to bind cooperatively to PEO backbones. This binding occurs when SDS concentrations are between 4.0 mM (critical aggregation concentration) and 16.5 mM (saturation concentration). As the concentration of SDS increases, the hydrodynamic radius of PEO unimers initially decreases and then increases. When the concentration of SDS is below the critical aggregation concentration (CAC) and above the C2 concentration, there is no binding interaction between SDS and PEO. Additionally, both Rh values remain unaffected by the SDS concentration. However, the relationship between the Rh (hydrodynamic radius) of PEO unimers and aggregates exhibits distinct patterns at different SDS concentrations, specifically between the critical aggregation concentration (CAC) and C2. In the case of unimeric PEO, SDS monomers link together with PEO chains by a process called polymer-induced micellization. As the concentration of SDS increases, the size of the resulting micelles decreases, while the PEO segments get hydrated and attach themselves to the surface of the SDS micelles. The size of the PEO chains increases, indicating this attachment. Nevertheless, as a result of the rise in the aggregation number and the electrostatic repulsion caused by the charged head groups of SDS, the hydrodynamic radius (Rh) of the large aggregates formed by SDS/PEO increases with the concentration of SDS, specifically between the critical aggregation concentration (CAC) and C2 [141, 142]. The researchers examine the impact of the drug aceclofenac (ACF) on the characteristics of three mixed micellar systems. The three systems consisted of pluronic L64 + F127 (nonionic-nonionic), pluronic L64 + CTAB (cetyltrimethylammonium bromide) (nonionic-cationic), and L64 + SDS (sodium dodecyl sulfate) (nonionic-anionic) combinations. The dynamic light scattering (DLS) measurement revealed that the addition of ACF increased the size of the individual micelle of pluronic L64 from 98 to 168 nm. The cationic mixed micelle with ACF had a size of 329 nm, while the anionic mixed micelle had a size of 291 nm. This indicates that the mixed micelles had higher entrapment efficiency compared to the single micelles. Pluronic block copolymers are recognized for their ability to exhibit phase behavioral characteristics. Therefore, they are widely employed in medication delivery applications. Their structure consists of both hydrophilic polyethylene oxide (PEO) and hydrophobic polypropylene oxide (PPO) pieces inside a single entity. The groups are organized in a certain structure, denoted as (PEO)n (PPO)m (PEO)n. They are accessible in a diverse spectrum of options, which vary based on different PO/EO ratios and molecular weights. Due to their amphiphilic properties, these substances have garnered interest from researchers in multiple disciplines, particularly for their potential to solubilize and deliver hydrophobic medicines [143, 144].

The alterations in the size of micelles caused by the medication aceclofenac in the mixed micellar system were quantified using dynamic light scattering with a zeta nanosizer, ZS (Marvern Panalytical, UK). The measurements were recorded at a temperature of $25.0 \pm 0.1°C$. A 4 mW He-Ne laser with a wavelength of 633 nm was employed as the light source. The samples were consistently kept at a scattering angle of 1730. Three different mixed micellar systems were investigated to observe the interplay of ACF. The solubilization of ACF was significantly increased in all three mixed micellar systems compared to their respective single micellar systems. The findings indicated that the ionic mixed micelles have more efficient interactions with ACF compared to the nonionic mixed micelles. In addition, the anionic mixed pair has a superior effect compared to the cationic pair. The results of the conductivity studies indicate that the interactions between ionic mixed micelles are thermodynamically possible, as evidenced by the counter ion binding constant and free energy of micellization. Furthermore, the higher negative value of the anionic mixed micelle (−27.8 kJ/mol) compared to the cationic mixed micelle (−25.4 kJ/mol) demonstrates that the former is more stable than the latter. The aforementioned study suggests that by using mixed micellar formulations, a surfactant can be adjusted to enhance the solubilization of ACF compared to single micellar solutions. Furthermore, when it comes to ACF, it is more advantageous to use an anionic mixed micellar system instead of a cationic mixture. These findings can be applied to other drug systems that repel water.

Li et al. produced mixed micelles using Pluronic P105 (P105) and poly(ethylene glycol)–phosphatidyl ethanolamine conjugate (PEG-PE). The research focused on examining the interaction between Pluronic and PEG-PE. CMC was measured using the laser dynamic light scattering method (DLLS) using DynaProTitan equipment (Wyatt Technology Corp., USA). A comparison was made between the changes in light intensity, and it was seen that there was a significant rise in the scattering intensity, which suggested the creation of micelles [145, 146]. The combination of Pluronic P105 and PEG-PE in mixed micelles exhibited synergistic effects. A negative value was seen for the parameter β, and the lowest CMC was achieved with an appropriate composition ratio. However, the synergy diminished as the quantity of PEG-PE rose. The observed behavior was ascribed to the adverse interactions between Pluronic and PEG-PE. The results demonstrated that the molar ratio significantly influenced the stability of mixed micelles. When a medicine contained in micelles is injected into the bloodstream, it becomes highly diluted, potentially leading to the dissociation of the micelles into individual molecules [147].

Despite the ongoing efforts of experts, cancer continues to cause millions of deaths worldwide. Currently, chemotherapy and radiotherapy remain the conventional approaches for treating malignancies. Nevertheless, the presence of inefficiencies and severe side effects typically results in decrease in patient compliance [148–152]. Therefore, clinical settings increasingly recognize innovative developments like photodynamic therapy (PDT) as the preferred option for tumor treatment. This approach offers the benefits of effectively eliminating specific tumor areas while avoiding harm to surrounding tissue with great accuracy in terms of both space and time [153, 154].

Nano-agents with a photoactivity "on-off" switch can protect against phototoxicity until they reach the target region, greatly enhancing photodynamic therapy. Nevertheless, the proportion of nano-agents that were concealed in living organisms varied and was directly linked to the light-induced toxicity caused by laser exposure. Consequently, the optimal timing for laser irradiation was uncertain to enhance the effectiveness of the treatment against tumors. The researchers connected low-molecular-weight chitosan and hydrophobic polymethylacrylamide derivatives using GSH cleavable 3, 3′-dithiodipropionic acid to create polymeric micelles (Ce6-CSPD). The nano-agent Ce6-CSPD/DOX can suppress both the photoactivity and fluorescence of the photosensitizer chlorin e6 (Ce6) and doxorubicin (DOX) in physiological conditions through homo-fluorescence resonance energy transfer (homoFRET). The dynamic fluorescence intensity can be used to guide the timing of laser irradiation, ensuring maximum effectiveness of anticancer therapy. This provides chances to assess the efficiency of chemo-photodynamic therapy in a timely and accurate manner. The development and metastasis of tumors were suppressed by the generation of deadly reactive oxygen species (ROS) with the use of near-infrared laser irradiation in the presence of PDT [155, 156]. Hence, nano-agents with the ability to turn the photoactivity "on" and "off" can suppress the photoactivity of the photodynamic reagent before it reaches the intended areas. This has significant potential in the field of cancer therapy [157–160]. Lu et al. [151] employed MOF nanoparticles as quenchers for photosensitizers. The MOF structures were capable of being efficiently dissolved by intracellular reducing agents.

Capillary electrophoresis (CE) uses laser-induced fluorescence (LIF) detection, a very sensitive method. The laser possesses the benefits of superb monochromaticity and high intensity, making it an ideal choice as an excitation light source for on-column detection in capillaries with reduced inner diameters. Curcuminoids can be detected with excellent sensitivity and specificity using laser-induced native fluorescence (LINF) due to their ability to emit native fluorescence when exposed to violet-blue light. A common 445 nm laser diode was utilized as an excitation source to create a confocal laser-induced fluorescence (LIF) detector. The LINF method that was created was tested and proven to be precise, linear, and accurate. A comprehensive CE-LIF system was built utilizing a commercially available 445 nm laser diode as the excitation source. The laser with a wavelength of 445 nm is ideal for exciting curcuminoids, and using LINF detection can successfully prevent interference from other components in the samples. Additionally, the combination of Trition-X 100/SDS mixed micelles can increase the sensitivity for analyzing curcuminoids by inducing fluorescence synergism [161, 162].

To monitor nanocarriers, numerous researchers utilize nanocarriers that are marked with radiotracers or containing near-infrared fluorescence (NIRF) dye. The study involves the synthesis of new amphiphilic copolymers, specifically methoxy poly(ethylene glycol) (mPEG)-cyanine-poly($\sum$ – caprolactone) (PCL) (mPEG-Cy-PCL). mPEGCy-PCL possesses the ability to conduct near-infrared fluorescence (NIRF) imaging, and photothermal treatment (PTT) on cancer cells, and act as self-assembling nanocarriers. Cy-based micelles can encapsulate doxorubicin (Doxo@Cy-micelle) and accomplish near-infrared fluorescence (NIRF) image-guided drug administration. Doxo@Cy-micelles are nanoscale micelles that promote the concentration of Doxo in tumor locations while reducing adverse effects. Doxo@Cy-micelles demonstrate a remarkable photothermal treatment (PTT) and a collaborative chemotherapeutic effect on cancer by the laser-induced release of Doxo from micelles, ultimately leading to reduced rates of cancer recurrence. The results indicate that micelles containing Cy are highly effective nanocarriers for near-infrared fluorescence (NIRF) imaging and the combined use of photothermal chemotherapy in cancer treatment (**Figure 7** shows an example application model). A new micelle, called Doxo@Cymicelle, was created using multifunctional copolymers, specifically mPEGCy-PCL. This micelle can be employed for both drug delivery and NIR fluorescence imaging-guided PTT concurrently [163]. The Doxo@ Cy-micelle is a type of nanomedicine that can increase the concentration of Doxo specifically in tumor sites while reducing the occurrence of unwanted side effects. Furthermore, upon laser irradiation, the Doxo@Cy-micelles not only absorb the laser energy but also transform it into heat, leading to the destruction of the micelles. The Doxo@Cymicelles achieved demonstrated the successful release of Doxo through remote laser activation, resulting in highly effective combined treatment of cancer using both chemotherapy and phototherapy. The findings suggest that the Empty@Cy-micelles were primarily responsible for the majority of the heat produced during laser irradiation. Organic dyes readily undergo photodegradation when exposed to laser irradiation. If Cy maintains

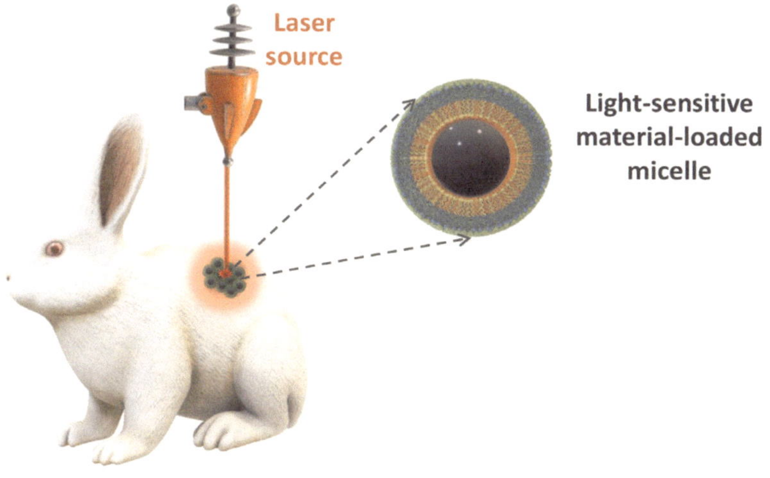

Figure 7.
Photothermal treatment on cancer cells with light-sensitive material-loaded micelle by laser light.

its optical properties throughout the treatment, it will be feasible to confirm the heat produced during laser exposure. The photodegradation of the Doxo@Cy-micelles was observed during laser irradiation by measuring the reduction in specific absorbance at a wavelength of 650 nm. Following a laser irradiation period of over 10 minutes, it was shown that the Doxo@Cy-micelles retained their optical characteristics. This indicates their ability to produce sufficient heat to eliminate cancer cells. Additionally, the polymer mPEG-Cy-PCL was found to effectively shield the Cy dyes from rapid photodegradation. Another possibility is that the self-assembled micelles had a higher concentration, causing the thermal radiation to be retained within the micelles. This could explain the observed high-energy efficiency and reduced heat dissipation [164]. Moreover, laser irradiation resulted in a higher release of Doxo from the Doxo@Cy-micelles. In the absence of laser irradiation, the release of Doxo from the Doxo@Cy-micelles was 31% after 12 hours, suggesting that the Doxo@Cy-micelles effectively reduce unwanted drug release. On the other hand, the Doxo@Cy-micelles exhibited significant enhancements in the rate at which the drug was released when exposed to laser irradiation (reaching up to 71.3%).

5. Conclusion

Nanomedicine, a new subspecialty of medicine, has emerged over the past decade. The study of nanoscale systems, namely polymeric micelles (PMs), which improve the solubility and stability of pharmaceuticals that are contained within them, is the primary emphasis of this research. PMs are distinguished by their core-shell architecture, which results in the possession of unique properties. These are suspensions of amphiphilic compounds that, when placed in water, develop a core-shell structure on their own. In contrast to the hydrophilic shell, which serves as a stabilizer, the core is hydrophobic and can contain hydrophobic drugs. In addition, because of their small size, PMs can be digested by the intestinal mucosa together with the medication, and they transport the medication through the bloodstream to the therapeutic site that is sought. Additionally, PMs improve the pharmacokinetic properties of the medication that is contained within them, demonstrate a significant capacity for transporting the medication, and are manufactured through a procedure that is reproducible, easy, and economical. In the last few decades, polymeric micelles have shown great promise as a breakthrough in the chemotherapy sector for the treatment of cancer. The hydrophilic outer shell and hydrophobic core of these micelles give them many benefits, such as their small size, ability to self-assembly, stability in both physical and chemical domains, compatibility with living things, ability to stay in the bloodstream for longer, and ability to transport and release a large amount of medication. Laser light has the potential to induce drug release from polymeric micelles. The disruption of the micelle structure through the integration of photosensitive elements, such as light-sensitive linkers or photothermal agents, enables the targeted release of the encapsulated drug from a designated site within the organism. It is possible to engineer polymeric micelles to contain fluorescent pigments or other imaging agents. Laser light can subsequently excite these agents, enabling imaging and sensing applications. Certain types of polymeric micelles are engineered to undergo photothermal therapy, a process in which they transform laser light into heat. These micelles generate localized hyperthermia, a therapeutic approach for cancer and other ailments, through their ability to accumulate selectively in target tissues and absorb laser light at specific wavelengths.

Acknowledgements

The authors thank to Karadeniz Technical University for their support. In addition, the authors also would like to express gratitude to Bilkent University UNAM for their kind hospitality. The authors would like to thank to Ibrahim ERDEMOGLU for their support.

Author details

Bulend Ortac[1], Saliha Mutlu[1,2], Ahmet Hakan Yilmaz[3] and Sevil Savaskan Yilmaz[1,2]*

1 Bilkent University, UNAM-National Nanotechnology Research Center and Institute of Materials Science and Nanotechnology, Ankara, Turkey

2 Faculty of Sciences, Department of Chemistry, Karadeniz Technical University, University Avenue, Trabzon, Turkey

3 Faculty of Sciences, Department of Physics, Karadeniz Technical University, University Avenue, Trabzon, Turkey

*Address all correspondence to: sevily@ktu.edu.tr

References

[1] Figueiras A, Domingues C, Jarak I, Santos AI, Parra A, Pais A, et al. New advances in biomedical application of polymeric micelles. Pharmaceutics. 2022;**14**:1700. DOI: 10.3390/pharmaceutics14081700

[2] Raja MAG. Various polymers in the development of polymeric micelles. In: Polymeric Micelles for Drug Delivery. United Kingdom: Elsevier; 2022. pp. 15-40. DOI: 10.1016/B978-0-323-89868-3.00024-0

[3] Jarak I, Pereira-Silva M, Santos AC, Veiga F, Cabral H, Figueiras A. Multifunctional polymeric micelle-based nucleic acid delivery: Current advances and future perspectives. Applied Materials Today. 2021;**25**:101217. DOI: 10.1016/j.apmt.2021.101217

[4] Bodratti AM, Alexandridis P. Amphiphilic block copolymers in drug delivery: Advances in formulation structure and performance. Expert Opinion on Drug Delivery. 2018;**2018**(15):1085-1104. DOI: 10.1080/17425247.2018.1529756

[5] Pérez SE, Haidar ZS. Polymeric micelles: Précis for past, present and future. In: Polymeric Micelles for Drug Delivery. United Kingdom: Elsevier; 2022. pp. 3-14

[6] Jones M-C, Leroux J-C. Polymeric micelles – A new generation of colloidal drug carriers. European Journal of Pharmaceutics and Biopharmaceutics. 1999;**48**:101-111. DOI: 10.1016/S0939-6411(99)00039-9

[7] Hwang D, Ramsey JD, Kabanov AV. Polymeric micelles for the delivery of poorly soluble drugs: From nanoformulation to clinical approval. Advanced Drug Delivery Reviews. 2020;**156**:80-118. DOI: 10.1016/j.addr.2020.09.009

[8] Gothwal A, Khan I, Gupta U. Polymeric micelles: Recent advancements in the delivery of anticancer drugs. Pharmaceutical Research. 2016;**33**:18-39. DOI: 10.1007/s11095-015-1784-1

[9] Nagarajan R, Barry M, Ruckenstein E. Unusual selectivity in solubilization by block copolymer micelles. Langmuir. 1986;**2**:210-215. DOI: 10.1021/la00068a017

[10] Arnida NN, Kanayama N, Jang W-D, Yamasaki Y, Kataoka K. PEGylated gene nanocarriers based on block catiomers bearing ethylenediamine repeating units directed to remarkable enhancement of photochemical transfection. Journal of Controlled Release. 2006;**115**:208-215. DOI: 10.1016/j.jconrel.2006.07.014

[11] Gref R, Domb A, Quellec P, Blunk T, Müller RH, Verbavatz JM, et al. The controlled intravenous delivery of drugs using PEG-coated sterically stabilized nanospheres. Advanced Drug Delivery Reviews. 1995;**16**:215-233. DOI: 10.1016/0169-409X(95)00026-4

[12] Hagan SA, Coombes AGA, Garnett MC, Dunn SE, Davies MC, Illum L, et al. Polylactide−poly(ethylene glycol) copolymers as drug delivery systems. 1. Characterization of water dispersible micelle-forming systems. Langmuir. 1996;**12**:2153-2161. DOI: 10.1021/la950649v

[13] Yokoyama M, Okano T, Sakurai Y, Suwa S, Kataoka K. Introduction of cisplatin into polymeric micelle. Journal of Controlled Release. 1996;**39**:351-356. DOI: 10.1016/0168-3659(95)00165-4

[14] Kwon GS, Okano T. Polymeric micelles as new drug carriers. Advanced Drug Delivery Reviews. 1996;**21**:107-116. DOI: 10.1016/S0169-409X(96)00401-2

[15] Kul D, Van Renterghem LM, Meier MAR, Strandman S, Tenhu H, Yılmaz SS, et al. Encapsulation and release by star-shaped block copolymers as unimolecular nanocontainers. Journal of Polymer Science Part A Polymer Chemistry. 2008;**46**:650-660. DOI: 10.1002/pola.22413

[16] Farkaš AZ, Popović K, Kumar D, Škorić D, Poša M. Regular solution theory regarding sodium cholate and hexadecyltrimethylammonium bromide or dodecyltrimethylammonium bromide binary mixed micelles. Journal of Molecular Liquids. 2023;**379**:121682. DOI: 10.1016/j.molliq.2023.121682

[17] Pilipović A, Ocokoljić M, Janev M, Poša M. The ternary mixed micelle of tween 20- sodium deoxycholate- sodium cholate: The molar excess thermodynamic potencials. The Journal of Chemical Thermodynamics. 2022;**167**:106695. DOI: 10.1016/j.jct.2021.106695

[18] Poša M. The Gibbs-Helmholtz equation and the enthalpy–entropy compensation (EEC) phenomenon in the formation of micelles in an aqueous solution of surfactants and the cloud point effect. Journal of Molecular Liquids. 2024;**396**:124109. DOI: 10.1016/j.molliq.2024.124109

[19] Kumar D, Poša M. Linear hydrophobic congeneric groups of bile acid anion derivatives based on the self-association (micellization) process and the phenomenon of enthalpy–entropy compensation. Journal of Molecular Liquids. 2023;**382**:121925. DOI: 10.1016/j.molliq.2023.121925

[20] Gao Z, Eisenberg A. A model of micellization for block copolymers in solutions. Macromolecules. 1993;**26**: 7353-7360. DOI: 10.1021/ma00078a035

[21] Cirin D, Posa M, Krstonosic V, Milanovic M. Conductometric study of sodium dodecyl sulfate - nonionic surfactant (triton X-100, tween 20, tween 60, tween 80 or tween 85) mixed micelles in aqueous solution. Hemijska industrija. 2012;**66**:21-28. DOI: 10.2298/HEMIND110612059C

[22] Puača G, Tepavčević V, Poša M. Interaction between triton X100 and Brij 58 in their binary mixed micelles: Micellization in aqueous solution and aqueous solution of Poloxamer 188 at the range of temperature T = (273.15–323.15) K. The Journal of Chemical Thermodynamics. 2022;**173**:106835. DOI: 10.1016/j.jct.2022.106835

[23] Poša M, Pilipović A. Effects of additives (methanol and NaCl) from aqueous surfactant solutions on the micellisation of sodium deoxycholate and sodium cholate binary mixture in the temperature interval T = (278.15–318.15) K: The molar excess Gibbs energy and the molar Gibbs e. The Journal of Chemical Thermodynamics. 2020;**150**:106179. DOI: 10.1016/j.jct.2020.106179

[24] Khan I, Gothwal A, Mishra G, Gupta U. Polymeric micelles. In: Jafar Mazumder MA, Sheardown H, Al-Ahmed A, editors. Functional Biopolymers. Cham, Switzerland: Springer International Publishing; 2019. pp. 73-102

[25] Kotta S, Aldawsari HM, Badr-Eldin SM, Nair AB, Yoonus Thajudeen K. Progress in polymeric micelles for drug delivery applications. Pharmaceutics. 2022;**14**:1636. DOI: 10.3390/pharmaceutics14081636

[26] Macha IJ, Ben-Nissan B, Vilchevskaya EN, Morozova AS, Abali BE, Müller WH, et al. Drug delivery from polymer-based

nanopharmaceuticals—An experimental study complemented by simulations of selected diffusion processes. Frontiers in Bioengineering and Biotechnology. 2019; **7**:1-14. DOI: 10.3389/fbioe.2019.00037

[27] Sung YK, Kim SW. Recent advances in polymeric drug delivery systems. Biomaterials Research. 2020;**24**:1-12. DOI: 10.1186/s40824-020-00190-7

[28] Liechty WB, Kryscio DR, Slaughter BV, Peppas NA. Polymers for drug delivery systems. Annual Review of Chemical and Biomolecular Engineering. 2010;**1**:149-173. DOI: 10.1146/annurev-chembioeng-073009-100847

[29] Almgren M. Mixed micelles and other structures in the solubilization of bilayer lipid membranes by surfactants. Biochimica et Biophysica Acta (BBA) - Biomembranes. 2000;**1508**:146-163. DOI: 10.1016/S0005-2736(00)00309-6

[30] Deshmukh AS, Chauhan PN, Noolvi MN, Chaturvedi K, Ganguly K, Shukla SS, et al. Polymeric micelles: Basic research to clinical practice. International Journal of Pharmaceutics. 2017;**532**:249-268. DOI: 10.1016/j.ijpharm.2017.09.005

[31] Kunitake T, Shinkai S. Catalysis by Micelles, membranes and other aqueous aggregates as models of enzyme action. Advances in Physical Organic Chemistry. 1980;**17**:435-487. DOI: 10.1016/S0065-3160(08)60131-8

[32] Kumar C, Balasubramanian D. Structural features of water-in-oil microemulsions. The Journal of Physical Chemistry. 1980;**84**:1895-1899. DOI: 10.1021/j100452a006

[33] Haider MS, Lübtow MM, Endres S, Forster S, Flegler VJ, Böttcher B, et al. Think beyond the Core: Impact of the hydrophilic Corona on drug solubilization using polymer micelles. ACS Applied Materials & Interfaces. 2020;**12**:24531-24543. DOI: 10.1021/acsami.9b22495

[34] Kulthe SS, Inamdar NN, Choudhari YM, Shirolikar SM, Borde LC, Mourya VK. Mixed micelle formation with hydrophobic and hydrophilic Pluronic block copolymers: Implications for controlled and targeted drug delivery. Colloids Surfaces B Biointerfaces. 2011; **88**:691-696. DOI: 10.1016/j.colsurfb.2011.08.002

[35] Manjappa AS, Kumbhar PS, Patil AB, Disouza JI, Patravale VB. Polymeric mixed micelles: Improving the anticancer efficacy of single-copolymer micelles. Critical Reviews in Therapeutic Drug Carrier Systems. 2019;**36**:1-58. DOI: 10.1615/CritRevTherDrugCarrierSyst.2018020481

[36] Wei Z, Hao J, Yuan S, Li Y, Juan W, Sha X, et al. Paclitaxel-loaded Pluronic P123/F127 mixed polymeric micelles: Formulation, optimization and in vitro characterization. International Journal of Pharmaceutics. 2009;**376**:176-185. DOI: 10.1016/j.ijpharm.2009.04.030

[37] Holmberg K, Jönsson B, Kronberg B, Lindman B. Surfactants and Polymers in Aqueous Solution. England: Wiley; 2002. pp. 1-527. ISBN: 0-471-49883-1

[38] Zhou Q, Rosen MJ. Molecular interactions of surfactants in mixed monolayers at the air/aqueous solution Interface and in mixed micelles in aqueous media: The regular solution approach. Langmuir. 2003;**19**:4555-4562. DOI: 10.1021/la020789m

[39] Kancharla S, Zoyhofski NA, Bufalini L, Chatelais BF, Alexandridis P. Association between nonionic amphiphilic polymer and ionic

surfactant in aqueous solutions: Effect of polymer hydrophobicity and Micellization. Polymers (Basel). 2020;**12**: 1831. DOI: 10.3390/polym12081831

[40] Ganguly R, Aswal VK, Hassan PA, Gopalakrishnan IK, Kulshreshtha SK. Effect of SDS on the self-assembly behavior of the PEO–PPO–PEO triblock copolymer (EO) 20 (PO) 70 (EO) 20. The Journal of Physical Chemistry. B. 2006;**110**:9843-9849. DOI: 10.1021/jp0607061

[41] Mishra J, Swain J, Mishra AK. Molecular level understanding of sodium dodecyl Sulfate (SDS) induced sol–gel transition of Pluronic F127 using Fisetin as a fluorescent molecular probe. The Journal of Physical Chemistry. B. 2018; **122**:181-193. DOI: 10.1021/acs.jpcb.7b10170

[42] Kumbhakar M. Aggregation of ionic surfactants to block copolymer assemblies: A simple fluorescence spectral study. The Journal of Physical Chemistry. B. 2007;**111**:14250-14255. DOI: 10.1021/jp077220k

[43] Kancharla S, Bedrov D, Tsianou M, Alexandridis P. Structure and composition of mixed micelles formed by nonionic block copolymers and ionic surfactants in water determined by small-angle neutron scattering with contrast variation. Journal of Colloid and Interface Science. 2022;**609**:456-468. DOI: 10.1016/j.jcis.2021.10.176

[44] Myhre S, Amann M, Willner L, Knudsen KD, Lund R. How detergents dissolve polymeric micelles: Kinetic pathways of hybrid micelle formation in SDS and block copolymer mixtures. Langmuir. 2020;**36**:12887-12899. DOI: 10.1021/acs.langmuir.0c02123

[45] Manzari MT, Shamay Y, Kiguchi H, Rosen N, Scaltriti M, Heller DA. Targeted drug delivery strategies for precision medicines. Nature Reviews Materials. 2021;**6**:351-370. DOI: 10.1038/s41578-020-00269-6

[46] Wells CM, Harris M, Choi L, Murali VP, Guerra FD, Jennings JA. Stimuli-responsive drug release from smart polymers. Journal of Functional Biomaterials. 2019;**10**:34. DOI: 10.3390/jfb10030034

[47] Saxena V, Hussain MD. Polymeric mixed micelles for delivery of curcumin to multidrug resistant ovarian cancer. Journal of Biomedical Nanotechnology. 2013;**9**:1146-1154. DOI: 10.1166/jbn.2013.1632

[48] Ebrahim Attia AB, Ong ZY, Hedrick JL, Lee PP, Ee PLR, Hammond PT, et al. Mixed micelles self-assembled from block copolymers for drug delivery. Current Opinion in Colloid & Interface Science. 2011;**16**:182-194. DOI: 10.1016/j.cocis.2010.10.003

[49] Cagel M, Tesan FC, Bernabeu E, Salgueiro MJ, Zubillaga MB, Moretton MA, et al. Polymeric mixed micelles as nanomedicines: Achievements and perspectives. European Journal of Pharmaceutics and Biopharmaceutics. 2017;**113**:211-228. DOI: 10.1016/j.ejpb.2016.12.019

[50] Gerardos AM, Balafouti A, Pispas S. Mixed copolymer micelles for nanomedicine. Nanomanufacturing. 2023;**3**:233-247. DOI: 10.3390/nanomanufacturing3020015

[51] Letellier P, Mayaffre A, Turmine M. Thoughts on the ideal behavior of mixed micelles and the appropriate application of regular solution theory (RST). Journal of Colloid and Interface Science. 2011; **354**:248-255. DOI: 10.1016/j.jcis.2010.09.082

[52] Schoenmaker L, Witzigmann D, Kulkarni JA, Verbeke R, Kersten G, Jiskoot W, et al. mRNA-lipid nanoparticle COVID-19 vaccines: Structure and stability. International Journal of Pharmaceutics. 2021;**601**: 120586. DOI: 10.1016/j.ijpharm.2021.120586

[53] Yang L, Gong L, Wang P, Zhao X, Zhao F, Zhang Z, et al. Recent advances in lipid nanoparticles for delivery of mRNA. Pharmaceutics. 2022;**14**:2682. DOI: 10.3390/pharmaceutics14122682

[54] Hur J, Chung AJ. Microfluidic and nanofluidic intracellular delivery. Advancement of Science. 2021;**8**:1-31. DOI: 10.1002/advs.202004595

[55] Kumar R, Santa Chalarca CF, Bockman MR, Van BC, Grimme CJ, Dalal RJ, et al. Polymeric delivery of therapeutic nucleic acids. Chemical Reviews. 2021;**121**:11527-11652. DOI: 10.1021/acs.chemrev.0c00997

[56] Rinkenauer AC, Schubert S, Traeger A, Schubert US. The influence of polymer architecture on in vitro pDNA transfection. Journal of Materials Chemistry B. 2015;**3**:7477-7493. DOI: 10.1039/C5TB00782H

[57] Aied A, Greiser U, Pandit A, Wang W. Polymer gene delivery: Overcoming the obstacles. Drug Discovery Today. 2013;**18**:1090-1098. DOI: 10.1016/j.drudis.2013.06.014

[58] Cabral H, Miyata K, Osada K, Kataoka K. Block copolymer micelles in nanomedicine applications. Chemical Reviews. 2018;**118**:6844-6892. DOI: 10.1021/acs.chemrev.8b00199

[59] Hanson MG, Grimme CJ, Santa Chalarca CF, Reineke TM. Cationic micelles outperform linear polymers for delivery of antisense oligonucleotides in serum: An exploration of polymer architecture, cationic moieties, and cell addition order. Bioconjugate Chemistry. 2022;**33**:2121-2131. DOI: 10.1021/acs.bioconjchem.2c00379

[60] Yousefpour MM, Yari KA. Polymeric micelles as mighty nanocarriers for cancer gene therapy: A review. Cancer Chemotherapy and Pharmacology. 2017;**79**:637-649. DOI: 10.1007/s00280-017-3273-1

[61] Mai Y, Eisenberg A. Self-assembly of block copolymers. Chemical Society Reviews. 2012;**41**:5969. DOI: 10.1039/c2cs35115c

[62] Pereira-Silva M, Jarak I, Santos AC, Veiga F, Figueiras A. Micelleplex-based nucleic acid therapeutics: From targeted stimuli-responsiveness to nanotoxicity and regulation. European Journal of Pharmaceutical Sciences. 2020;**153**: 105461. DOI: 10.1016/j.ejps.2020.105461

[63] Jiang Y, Reineke TM, Lodge TP. Complexation of DNA with cationic copolymer micelles: Effects of DNA length and topology. Macromolecules. 2018;**51**:1150-1160. DOI: 10.1021/acs.macromol.7b02201

[64] Tan Z, Jiang Y, Zhang W, Karls L, Lodge TP, Reineke TM. Polycation architecture and assembly direct successful gene delivery: Micelleplexes outperform polyplexes via optimal DNA packaging. Journal of the American Chemical Society. 2019;**141**:15804-15817. DOI: 10.1021/jacs.9b06218

[65] Wang H, Ding S, Zhang Z, Wang L, You Y. Cationic micelle: A promising nanocarrier for gene delivery with high transfection efficiency. The Journal of Gene Medicine. 2019;**21**:1-16, 1-26. DOI: 10.1002/jgm.3101

[66] Hu C-MJ, Fang RH, Luk BT, Zhang L. Polymeric nanotherapeutics: Clinical development and advances in stealth functionalization strategies. Nanoscale. 2014;**6**:65-75. DOI: 10.1039/C3NR05444F

[67] Schöttler S, Landfester K, Mailänder V. Controlling the stealth effect of nanocarriers through understanding the protein Corona. Angewandte Chemie International Edition. 2016;**55**:8806-8815. DOI: 10.1002/anie.201602233

[68] Fang Y, Xue J, Gao S, Lu A, Yang D, Jiang H, et al. Cleavable PEGylation: A strategy for overcoming the "PEG dilemma" in efficient drug delivery. Drug Delivery. 2017;**24**:22-32. DOI: 10.1080/10717544.2017.1388451

[69] Zhang Y, Satterlee A, Huang L. In vivo gene delivery by nonviral vectors: Overcoming hurdles? Molecular Therapy. 2012;**20**:1298-1304. DOI: 10.1038/mt.2012.79

[70] Ishihara T, Maeda T, Sakamoto H, Takasaki N, Shigyo M, Ishida T, et al. Evasion of the accelerated blood clearance phenomenon by coating of nanoparticles with various hydrophilic polymers. Biomacromolecules. 2010;**11**:2700-2706. DOI: 10.1021/bm100754e

[71] Abu Lila AS, Kiwada H, Ishida T. The accelerated blood clearance (ABC) phenomenon: Clinical challenge and approaches to manage. Journal of Controlled Release. 2013;**172**:38-47. DOI: 10.1016/j.jconrel.2013.07.026

[72] Yao X, Qi C, Sun C, Huo F, Jiang X. Poly(ethylene glycol) alternatives in biomedical applications. Nano Today. 2023;**48**:101738. DOI: 10.1016/j.nantod.2022.101738

[73] Kierstead PH, Okochi H, Venditto VJ, Chuong TC, Kivimae S, Fréchet JMJ, et al. The effect of polymer backbone chemistry on the induction of the accelerated blood clearance in polymer modified liposomes. Journal of Controlled Release. 2015;**213**:1-9. DOI: 10.1016/j.jconrel.2015.06.023

[74] Bludau H, Czapar AE, Pitek AS, Shukla S, Jordan R, Steinmetz NF. POxylation as an alternative stealth coating for biomedical applications. European Polymer Journal. 2017;**8**:679-688. DOI: 10.1016/j.eurpolymj.2016.10.041

[75] Grube M, Cinar G, Schubert US, Nischang I. Incentives of using the hydrodynamic invariant and sedimentation parameter for the study of naturally- and synthetically-based macromolecules in solution. Polymers (Basel). 2020;**12**:277. DOI: 10.3390/polym12020277

[76] Hu Y, Hou Y, Wang H, Lu H. Polysarcosine as an alternative to PEG for therapeutic protein conjugation. Bioconjugate Chemistry. 2018;**29**:2232-2238. DOI: 10.1021/acs.bioconjchem.8b00237

[77] Nogueira SS, Schlegel A, Maxeiner K, Weber B, Barz M, Schroer MA, et al. Polysarcosine-functionalized lipid nanoparticles for therapeutic mRNA delivery. ACS Applied Nano Materials. 2020;**3**:10634-10645. DOI: 10.1021/acsanm.0c01834

[78] Lin T, Zhu T, Xun Y, Tao Y, Yang Y, Xie J, et al. A novel drug delivery system of mixed micelles based on poly(ethylene glycol)-poly(lactide) and poly(ethylene glycol)-poly(ε-caprolactone) for gambogenic acid. The Kaohsiung Journal of Medical Sciences. 2019;**35**:757-764. DOI: 10.1002/kjm2.12110

[79] Nakashima K, Bahadur P. Aggregation of water-soluble block

copolymers in aqueous solutions: Recent trends. Advances in Colloid and Interface Science. 2006;**123–126**:75-96. DOI: 10.1016/j.cis.2006.05.016

[80] Kwon SH, Kim SY, Ha KW, Kang MJ, Huh JS, Tae JI, et al. Pharmaceutical evaluation of genistein-loaded pluronic micelles for oral delivery. Archives of Pharmacal Research. 2007;**30**:1138-1143. DOI: 10.1007/BF02980249

[81] Wang Y, Li Y, Zhang L, Fang X. Pharmacokinetics and biodistribution of paclitaxel-loaded pluronic P105 polymeric micelles. Archives of Pharmacal Research. 2008;**31**:530-538. DOI: 10.1007/s12272-001-1189-2

[82] Ezhilrani VC, Karunanithi P, Sarangi B, Joshi RG, Dash S. Hydrophilic-hydrophilic mixed micellar system: Effect on solubilization of drug. SN Applied Sciences. 2021;**3**:371. DOI: 10.1007/s42452-021-04250-y

[83] Razavi L, Raissi H, Farzad F. Validation of an MD simulation approach for electrical field responsive micelles and their application in drug delivery. Scientific Reports. 2023;**13**: 2665. DOI: 10.1038/s41598-023-29835-y

[84] Brooks BR, Brooks CL, Mackerell AD, Nilsson L, Petrella RJ, Roux B, et al. CHARMM: The biomolecular simulation program. Journal of Computational Chemistry. 2009;**30**:1545-1614. DOI: 10.1002/jcc.21287

[85] Lee S, Tran A, Allsopp M, Lim JB, Hénin J, Klauda JB. CHARMM36 united atom chain model for lipids and surfactants. The Journal of Physical Chemistry. B. 2014;**118**:547-556. DOI: 10.1021/jp410344g

[86] Razavi L, Raissi H, Farzad F. Assessment of the effect of external and internal triggers on adsorption and release of paclitaxel from the PEI functionalized silicene nanosheet: A molecular dynamic simulation. Journal of Molecular Graphics & Modelling. 2021;**106**:107930. DOI: 10.1016/j.jmgm.2021.107930

[87] Lin Y, Pan D, Li J, Zhang L, Shao X. Application of Berendsen barostat in dissipative particle dynamics for nonequilibrium dynamic simulation. The Journal of Chemical Physics. 2017;**146**:124108. DOI: 10.1063/1.4978807

[88] Razavi L, Raissi H, Farzad F. Insights into glyphosate removal efficiency using a new 2D nanomaterial. RSC Advances. 2022;**12**:10154-10161. DOI: 10.1039/D2RA00385F

[89] Matsumoto R, Thompson MW, Cummings PT. Ion pairing controls physical properties of ionic liquid-solvent mixtures. The Journal of Physical Chemistry. B. 2019;**123**:9944-9955. DOI: 10.1021/acs.jpcb.9b08509

[90] Kasturiarachi AB. Leap-frogging Newton's method. International Journal of Mathematical Education in Science and Technology. 2002;**33**:521-527. DOI: 10.1080/00207390210131786

[91] Cheatham TEI, Miller JL, Fox T, Darden TA, Kollman PA. Molecular dynamics simulations on solvated biomolecular systems: The particle mesh Ewald method leads to stable trajectories of DNA, RNA, and proteins. Journal of the American Chemical Society. 1995; **117**:4193-4194. DOI: 10.1021/ja00119a045

[92] Hess B, Bekker H, Berendsen HJC, Fraaije JGEM. LINCS: A linear constraint solver for molecular simulations. Journal of Computational Chemistry. 1997;**18**: 1463-1472. DOI: 10.1002/(SICI)1096-

987X(199709)18:12<1463::AID-JCC4>3.0.CO;2-H

[93] Humphrey W, Dalke A, Schulten K. VMD: Visual molecular dynamics. Journal of Molecular Graphics. 1996;**14**: 33-38. DOI: 10.1016/0263-7855(96) 00018-5

[94] Alasiri H. Determining critical micelle concentrations of surfactants based on viscosity calculations from coarse-grained molecular dynamics simulations. Energy & Fuels. 2019;**33**: 2408-2412. DOI: 10.1021/acs.energyfuels.8b04228

[95] Meng X, Liu J, Yu X, Li J, Lu X, Shen T. Pluronic F127 and D-α-tocopheryl polyethylene glycol succinate (TPGS) mixed micelles for targeting drug delivery across the blood brain barrier. Scientific Reports. 2017;**7**:2964. DOI: 10.1038/s41598-017-03123-y

[96] Kim SC, Kim DW, Shim YH, Bang JS, Oh HS, Kim SW, et al. In vivo evaluation of polymeric micellar paclitaxel formulation: Toxicity and efficacy. Journal of Controlled Release. 2001;**72**:191-202. DOI: 10.1016/S0168-3659(01)00275-9

[97] Gupta A. Ketorolac entrapped in polymeric micelles: Preparation, characterisation and ocular anti-inflammatory studies. International Journal of Pharmaceutics. 2000;**209**:1-14. DOI: 10.1016/S0378-5173(00) 00508-1

[98] Zhao D, Wu J, Li C, Zhang H, Li Z, Luan Y. Precise ratiometric loading of PTX and DOX based on redox-sensitive mixed micelles for cancer therapy. Colloids Surfaces B Biointerfaces. 2017; **155**:51-60

[99] Salama AH, Shamma RN. Tri/tetra-block co-polymeric nanocarriers as a potential ocular delivery system of lornoxicam: In-vitro characterization, and in-vivo estimation of corneal permeation. International Journal of Pharmaceutics. 2015;**492**:28-39. DOI: 10.1016/j.ijpharm.2015.07.010

[100] Raza K, Kumar N, Misra C, Kaushik L, Guru SK, Kumar P, et al. Dextran-PLGA-loaded docetaxel micelles with enhanced cytotoxicity and better pharmacokinetic profile. International Journal of Biological Macromolecules. 2016;**88**:206-212. DOI: 10.1016/j.ijbiomac.2016.03.064

[101] Ribeiro A, Sandez-Macho I, Casas M, Alvarez-Pérez S, Alvarez-Lorenzo C, Concheiro A. Poloxamine micellar solubilization of α-tocopherol for topical ocular treatment. Colloids Surfaces B Biointerfaces. 2013;**103**:550-557. DOI: 10.1016/j.colsurfb.2012.10.055

[102] Yoo HS, Lee EA, Park TG. Doxorubicin-conjugated biodegradable polymeric micelles having acid-cleavable linkages. Journal of Controlled Release. 2002;**82**:17-27. DOI: 10.1016/S0168-3659 (02)00088-3

[103] Kadam Y, Yerramilli U, Bahadur A. Solubilization of poorly water-soluble drug carbamezapine in Pluronic® micelles: Effect of molecular characteristics, temperature and added salt on the solubilizing capacity. Colloids Surfaces B Biointerfaces. 2009;**72**:141-147. DOI: 10.1016/j.colsurfb.2009.03.027

[104] Yoo HS, Park TG. Folate receptor targeted biodegradable polymeric doxorubicin micelles. Journal of Controlled Release. 2004;**96**:273-283. DOI: 10.1016/j.jconrel.2004.02.003

[105] Pepić I, Hafner A, Lovrić J, Pirkić B, Filipović-Grččić J. A nonionic surfactant/chitosan micelle system in an innovative eye drop formulation. Journal of

Pharmaceutical Sciences. 2010;**99**:4317-4325. DOI: 10.1002/jps.22137

[106] Wang Z, Liu H, Shu X, Zheng L, Chen L. A reduction-degradable polymer prodrug for cisplatin delivery: Preparation, in vitro and in vivo evaluation. Colloids Surfaces B Biointerfaces. 2015;**136**:160-167. DOI: 10.1016/j.colsurfb.2015.09.010

[107] Chopra P, Hao J, Li SK. Influence of drug lipophilicity on drug release from sclera after Iontophoretic delivery of mixed micellar carrier system to human sclera. Journal of Pharmaceutical Sciences. 2013;**102**:480-488. DOI: 10.1002/jps.23370

[108] Bui QN, Li Y, Jang M-S, Huynh DP, Lee JH, Lee DS. Redox- and pH-sensitive polymeric micelles based on poly(β-amino ester)-grafted disulfide methylene oxide poly(ethylene glycol) for anticancer drug delivery. Macromolecules. 2015;**48**:4046-4054. DOI: 10.1021/acs.macromol.5b00423

[109] Li Q, Lai KL, Chan PS, Leung SC, Li HY, Fang Y, et al. Micellar delivery of dasatinib for the inhibition of pathologic cellular processes of the retinal pigment epithelium. Colloids Surfaces B Biointerfaces. 2016;**140**:278-286. DOI: 10.1016/j.colsurfb.2015.12.053

[110] Taha EI, Badran MM, El-Anazi MH, Bayomi MA, El-Bagory IM. Role of pluronic F127 micelles in enhancing ocular delivery of ciprofloxacin. Journal of Molecular Liquids. 2014;**199**:251-256. DOI: 10.1016/j.molliq.2014.09.021

[111] Tong Y-C, Chang S-F, Kao WW-Y, Liu C-Y, Liaw J. Polymeric micelle gene delivery of bcl-xL via eye drop reduced corneal apoptosis following epithelial debridement. Journal of Controlled Release. 2010;**147**:76-83. DOI: 10.1016/j.jconrel.2010.06.006

[112] Jaiswal M, Kumar M, Pathak K. Zero order delivery of itraconazole via polymeric micelles incorporated in situ ocular gel for the management of fungal keratitis. Colloids Surfaces B Biointerfaces. 2015;**130**:23-30. DOI: 10.1016/j.colsurfb.2015.03.059

[113] Binkhathlan Z, Ali R, Alomrani AH, Abul Kalam M, Alshamsan A, Lavasanifar A. Role of polymeric micelles in ocular drug delivery: An overview of decades of research. Molecular Pharmaceutics. 2023;**20**:5359-5382. DOI: 10.1021/acs.molpharmaceut.3c00598

[114] Figueroa-Ochoa EB, Villar-Alvarez EM, Cambón A, Mistry D, Llovo J, Attwood D, et al. Lenghty reverse poly(butylene oxide)-poly(ethylene oxide)-poly(butylene oxide) polymeric micelles and gels for sustained release of antifungal drugs. International Journal of Pharmaceutics. 2016;**510**:17-29. DOI: 10.1016/j.ijpharm.2016.06.013

[115] Civiale C, Licciardi M, Cavallaro G, Giammona G, Mazzone MG. Polyhydroxyethylaspartamide-based micelles for ocular drug delivery. International Journal of Pharmaceutics. 2009;**378**:177-186. DOI: 10.1016/j.ijpharm.2009.05.028

[116] Wu Y, Zhang Y, Zhang W, Sun C, Wu J, Tang J. Reversing of multidrug resistance breast cancer by co-delivery of P-gp siRNA and doxorubicin via folic acid-modified core-shell nanomicelles. Colloids Surfaces B Biointerfaces. 2016;**138**:60-69. DOI: 10.1016/j.colsurfb.2015.11.041

[117] Lu C, Yoganathan RB, Kociolek M, Allen C. Hydrogel containing silica shell cross-linked micelles for ocular drug delivery. Journal of Pharmaceutical Sciences. 2013;**102**:627-637. DOI: 10.1002/jps.23390

[118] Xiao Y, Wang S, Zong Q, Yin Z. Co-delivery of metformin and paclitaxel via folate-modified pH-sensitive micelles for enhanced anti-tumor efficacy. AAPS PharmSciTech. 2018;**19**:2395-2406. DOI: 10.1208/s12249-018-1070-8

[119] Di Tommaso C, Bourges J-L, Valamanesh F, Trubitsyn G, Torriglia A, Jeanny J-C, et al. Novel micelle carriers for cyclosporin a topical ocular delivery: In vivo cornea penetration, ocular distribution and efficacy studies. European Journal of Pharmaceutics and Biopharmaceutics. 2012;**81**:257-264. DOI: 10.1016/j.ejpb.2012.02.014

[120] Guo L-Y, Yan S-Z, Tao X, Yang Q, Li Q, Wang T-S, et al. Evaluation of hypocrellin A-loaded lipase sensitive polymer micelles for intervening methicillin-resistant Staphylococcus Aureus antibiotic-resistant bacterial infection. Materials Science and Engineering: C. 2020;**106**:110230. DOI: 10.1016/j.msec.2019.110230

[121] Wang G, Nie Q, Zang C, Zhang B, Zhu Q, Luo G, et al. Self-assembled thermoresponsive nanogels prepared by reverse micelle → positive micelle method for ophthalmic delivery of Muscone, a poorly water-soluble drug. Journal of Pharmaceutical Sciences. 2016;**105**:2752-2759. DOI: 10.1016/j.xphs.2016.02.014

[122] Gener P, Montero S, Xandri-Monje H, Díaz-Riascos ZV, Rafael D, Andrade F, et al. Zileuton™ loaded in polymer micelles effectively reduce breast cancer circulating tumor cells and intratumoral cancer stem cells. Nanomedicine: Nanotechnology, Biology and Medicine. 2020;**24**:102106. DOI: 10.1016/j.nano.2019.102106

[123] Ma F, Nan K, Lee S, Beadle JR, Hou H, Freeman WR, et al. Micelle formulation of hexadecyloxypropyl-cidofovir (HDP-CDV) as an intravitreal long-lasting delivery system. European Journal of Pharmaceutics and Biopharmaceutics. 2015;**89**:271-279. DOI: 10.1016/j.ejpb.2014.12.010

[124] Supasena W, Muangnoi C, Thaweesest W, Songkram C, Ueda K, Higashi K, et al. Enhanced antipsoriatic activity of mycophenolic acid against the TNF-α-induced HaCaT cell proliferation by conjugated poloxamer micelles. Journal of Pharmaceutical Sciences. 2020;**109**:1153-1160. DOI: 10.1016/j.xphs.2019.11.010

[125] Khan I, Gothwal A, Sharma AK, Kesharwani P, Gupta L, Iyer AK, et al. PLGA nanoparticles and their versatile role in anticancer drug delivery. Critical Reviews in Therapeutic Drug Carrier Systems. 2016;**33**:159-193. DOI: 10.1615/CritRevTherDrugCarrierSyst.2016015273

[126] Jin Z, Nguyen KT, Go G, Kang B, Min H-K, Kim S-J, et al. Multifunctional nanorobot system for active therapeutic delivery and synergistic chemo-photothermal therapy. Nano Letters. 2019;**19**:8550-8564. DOI: 10.1021/acs.nanolett.9b03051

[127] Li N, Sun Q, Yu Z, Gao X, Pan W, Wan X, et al. Nuclear-targeted photothermal therapy prevents cancer recurrence with near-infrared triggered copper Sulfide nanoparticles. ACS Nano. 2018;**12**:5197-5206. DOI: 10.1021/acsnano.7b06870

[128] Santibanez JF, Villar VH, Echeverria C. Current and future cancer chemoprevention strategies. Pharmaceutics. 2023;**15**:1543. DOI: 10.3390/pharmaceutics15051543

[129] Bholakant R, Dong B, Zhou X, Huang X, Zhao C, Huang D, et al. Multi-functional polymeric micelles for

chemotherapy-based combined cancer therapy. Journal of Materials Chemistry B. 2021;**9**:8718-8738. DOI: 10.1039/D1TB01771C

[130] Guo K. Application of polymeric micelles for cancer treatment: A review. Highlights in Science Engineering and Technology. 2023;**45**:160-169. DOI: 10.54097/hset.v45i.7361

[131] Majumder NG, Das N, Das SK. Polymeric micelles for anticancer drug delivery. Therapeutic Delivery. 2020;**11**: 613-635. DOI: 10.4155/tde-2020-0008

[132] Afsharzadeh M, Hashemi M, Mokhtarzadeh A, Abnous K, Ramezani M. Recent advances in co-delivery systems based on polymeric nanoparticle for cancer treatment. Artificial Cells, Nanomedicine, and Biotechnology. 2018; **46**:1095-1110. DOI: 10.1080/21691401.2017.1376675

[133] Gibot L, Lemelle A, Till U, Moukarzel B, Mingotaud A-F, Pimienta V, et al. Polymeric micelles encapsulating photosensitizer: Structure/photodynamic therapy efficiency relation. Biomacromolecules. 2014;**15**: 1443-1455. DOI: 10.1021/bm5000407

[134] Jin G-W, Rejinold NS, Choy J-H. Multifunctional polymeric micelles for cancer therapy. Polymers (Basel). 2022; **14**:4839. DOI: 10.3390/polym14224839

[135] Wan D, Li C, Pan J. Polymeric micelles with reduction-responsive function for targeted cancer chemotherapy. ACS Applied Bio Materials. 2020;**3**:1139-1146. DOI: 10.1021/acsabm.9b01070

[136] Al Zaki A, Joh D, Cheng Z, De Barros ALB, Kao G, Dorsey J, et al. Gold-loaded polymeric micelles for computed tomography-guided radiation therapy treatment and radiosensitization. ACS Nano. 2014;**8**:104-112. DOI: 10.1021/nn405701q

[137] Tomoda K, Tam YT, Cho H, Buehler D, Kozak KR, Kwon GS. Triolimus: A multi-drug loaded polymeric micelle containing paclitaxel, 17-AAG, and rapamycin as a novel radiosensitizer. Macromolecular Bioscience. 2017;**17**:1600194. DOI: 10.1002/mabi.201600194

[138] Yang X, Chen X, Wang Y, Xu G, Yu L, Ding J. Sustained release of lipophilic gemcitabine from an injectable polymeric hydrogel for synergistically enhancing tumor chemoradiotherapy. Chemical Engineering Journal. 2020;**396**: 125320. DOI: 10.1016/j.cej.2020.125320

[139] Liu CL. Investigation of the interaction between simple micelles and mixed micelles in taurocholate–lecithin solutions by laser light scattering. The Journal of Physical Chemistry. B. 1998; **102**:5727-5732. DOI: 10.1021/jp980901i

[140] Xie D, Xu K, Bai R, Zhang G. Structural evolution of mixed micelles due to interchain complexation and segregation investigated by laser light scattering. The Journal of Physical Chemistry. B. 2007;**111**:778-781. DOI: 10.1021/jp066438o

[141] Dai S, Tam KC. Laser light scattering and isothermal titration calorimetric studies of poly(ethylene oxide) aqueous solution in presence of sodium dodecyl sulfate. Journal of Colloid and Interface Science. 2005;**292**: 79-85. DOI: 10.1016/j.jcis.2005.05.079

[142] Sheelarani B, Paul Raj E, Joshi RG, Dash S. Effect of drug aceclofenac on physicochemical properties of mixed micellar systems. SN Applied Sciences. 2022;**4**:1-10. DOI: 10.1007/s42452-022-05055-3

[143] Batrakova EV, Kabanov AV. Pluronic block copolymers: Evolution of drug delivery concept from inert nanocarriers to biological response modifiers. Journal of Controlled Release. 2008;**130**:98-106. DOI: 10.1016/j.jconrel.2008.04.013

[144] Kadam Y, Yerramilli U, Bahadur A, Bahadur P. Micelles from PEO–PPO–PEO block copolymers as nanocontainers for solubilization of a poorly water soluble drug hydrochlorothiazide. Colloids Surfaces B Biointerfaces. 2011;**83**:49-57. DOI: 10.1016/j.colsurfb.2010.10.041

[145] Li L, Tan YB. Preparation and properties of mixed micelles made of Pluronic polymer and PEG-PE. Journal of Colloid and Interface Science. 2008; **317**:326-331. DOI: 10.1016/j.jcis.2007.09.053

[146] Alexandridis P, Alan Hatton T. Poly (ethylene oxide) poly(propylene oxide) poly(ethylene oxide) block copolymer surfactants in aqueous solutions and at interfaces: Thermodynamics, structure, dynamics, and modeling. Colloids and Surfaces A: Physicochemical and Engineering Aspects. 1995;**96**:1-46. DOI: 10.1016/0927-7757(94)03028-X

[147] Zhu Y, Yu F, Tan Y, Wen L, Li Y, Yuan H, et al. Guiding appropriate timing of laser irradiation by polymeric micelles for maximizing chemo-photodynamic therapy. International Journal of Nanomedicine. 2020;**15**:6531-6543. DOI: 10.2147/IJN.S256477

[148] Shanmugam V, Selvakumar S, Yeh C-S. Near-infrared light-responsive nanomaterials in cancer therapeutics. Chemical Society Reviews. 2014;**43**: 6254-6287. DOI: 10.1039/C4CS00011K

[149] Gai S, Yang G, Yang P, He F, Lin J, Jin D, et al. Recent advances in functional nanomaterials for light–triggered cancer therapy. Nano Today. 2018;**19**:146-187. DOI: 10.1016/j.nantod.2018.02.010

[150] Wang T, Wang D, Yu H, Wang M, Liu J, Feng B, et al. Intracellularly acid-switchable multifunctional micelles for combinational photo/chemotherapy of the drug-resistant tumor. ACS Nano. 2016;**10**:3496-3508. DOI: 10.1021/acsnano.5b07706

[151] Lu K, He C, Guo N, Chan C, Ni K, Weichselbaum RR, et al. Chlorin-based nanoscale metal–organic framework systemically rejects colorectal cancers via synergistic photodynamic therapy and checkpoint blockade immunotherapy. Journal of the American Chemical Society. 2016;**138**:12502-12510. DOI: 10.1021/jacs.6b06663

[152] Ekladious I, Colson YL, Grinstaff MW. Polymer–drug conjugate therapeutics: Advances, insights and prospects. Nature Reviews Drug Discovery. 2019;**18**:273-294. DOI: 10.1038/s41573-018-0005-0

[153] Zhang Z, Wang R, Huang X, Luo R, Xue J, Gao J, et al. Self-delivered and self-monitored chemo-photodynamic nanoparticles with light-triggered synergistic antitumor therapies by downregulation of HIF-1α and depletion of GSH. ACS Applied Materials & Interfaces. 2020;**12**:5680-5694. DOI: 10.1021/acsami.9b23325

[154] Xu M, Yang G, Bi H, Xu J, Dong S, Jia T, et al. An intelligent nanoplatform for imaging-guided photodynamic/photothermal/chemo-therapy based on upconversion nanoparticles and CuS integrated black phosphorus. Chemical Engineering Journal. 2020; **382**:122822. DOI: 10.1016/j.cej.2019.122822

[155] Ethirajan M, Chen Y, Joshi P, Pandey RK. The role of porphyrin chemistry in tumor imaging and photodynamic therapy. Chemical Society Reviews. 2011;**40**:340-362. DOI: 10.1039/B915149B

[156] Ding Y-F, Li S, Liang L, Huang Q, Yuwen L, Yang W, et al. Highly biocompatible chlorin e6-loaded chitosan nanoparticles for improved photodynamic cancer therapy. ACS Applied Materials & Interfaces. 2018;**10**: 9980-9987. DOI: 10.1021/acsami.8b01522

[157] Wang Y, Zhou K, Huang G, Hensley C, Huang X, Ma X, et al. A nanoparticle-based strategy for the imaging of a broad range of tumours by nonlinear amplification of microenvironment signals. Nature Materials. 2014;**13**:204-212. DOI: 10.1038/nmat3819

[158] Wan Y, Lu G, Zhang J, Wang Z, Li X, Chen R, et al. A biocompatible free radical nanogenerator with real-time monitoring capability for high performance sequential hypoxic tumor therapy. Advanced Functional Materials. 2019;**29**:1903436. DOI: 10.1002/adfm.201903436

[159] Cheng P, Pu K. Activatable phototheranostic materials for imaging-guided cancer therapy. ACS Applied Materials & Interfaces. 2020;**12**:5286-5299. DOI: 10.1021/acsami.9b15064

[160] Yu L, Chen Y, Wu M, Cai X, Yao H, Zhang L, et al. "Manganese extraction" strategy enables tumor-sensitive biodegradability and theranostics of nanoparticles. Journal of the American Chemical Society. 2016;**138**:9881-9894. DOI: 10.1021/jacs.6b04299

[161] Wu C, Wang W, Quan F, Chen P, Qian J, Zhou L, et al. Sensitive analysis of curcuminoids via micellar electrokinetic chromatography with laser-induced native fluorescence detection and mixed micelles-induced fluorescence synergism. Journal of Chromatography. A. 2018;**1564**:207-213. DOI: 10.1016/j.chroma.2018.06.012

[162] Chen Y, Peng C, Lee P, Tsai M, Lin C, Shih Y, et al. Traceable self-assembly of laser-triggered cyanine-based micelle for synergistic therapeutic effect. Advanced Healthcare Materials. 2015;**4**: 892-902. DOI: 10.1002/adhm.201400729

[163] Croissant JG, Mongin O, Hugues V, Blanchard-Desce M, Cattoën X, Wong Chi Man M, et al. Influence of the synthetic method on the properties of two-photon-sensitive mesoporous silica nanoparticles. Journal of Materials Chemistry B. 2015;**3**:5182-5188. DOI: 10.1039/c5tb00787a

[164] Wan Z, Mao H, Guo M, Li Y, Zhu A, Yang H, et al. Highly efficient hierarchical micelles integrating photothermal therapy and singlet oxygen-synergized chemotherapy for cancer eradication. Theranostics. 2014;**4**: 399-411. DOI: 10.7150/thno.8171

Chapter 2

Micellar Polymer Flooding

Abdelaziz El-Hoshoudy

Abstract

This chapter discusses one of the crucial chemical-enhanced oil recovery systems associated with a combination of micellar and polymer flooding during the tertiary recovery process. Generally, polymer flooding relies on increasing displacing fluid viscosity to improve mobility ratio and sweeping efficiency. on the other hand, micellar flooding aims to decrease O/W interfacial tension, so improve displacement efficiency. The combination of both techniques and investigating their synergistic effect during the flooding process is a crucial issue that helps petroleum engineers assign the optimum flooding technique through reservoir management. The current trends and updated technologies in this field as well as previous literature will be discussed highlighting the sides of their advantages and limitations.

Keywords: enhanced oil recovery (EOR), micellar flooding, polymer flooding, wettability alteration, interfacial tension (IFT)

1. Introduction

Petroleum reservoirs yield approximately 33–50% of the initial oil reserves, resulting in substantial quantities of remaining oil that become the focal point of various Enhanced Oil Recovery (EOR) methods [1, 2]. EOR methods aim to enhance reservoir performance by aiding the movement of oil through the porous rock formation [3]. The utilization of chemical flooding techniques in oil reservoirs has proven to be a highly effective approach for improving oil recovery from depleted reservoirs operating at reduced pressures following secondary water flooding. Chemical Enhanced Oil Recovery (EOR) methods involve the injection of substances not naturally present in the reservoir to augment oil displacement [4]. Chemical flooding encompasses a range of displacement agents, including polymers, surfactants, and the combination thereof referred to as micellar/polymer flooding. Furthermore, it encompasses the use of acids, foaming agents, and solvents. Additionally, the combined approach of alkaline/surfactant/polymer (ASP) flooding has gained notable attention in various petroleum applications, including EOR, oil remediation, and the dispersion of oil spills [5]. Surfactants and polymers are the primary constituents employed in chemical flooding, hence, chemical flooding is commonly referred to interchangeably as micellar/polymer flooding. Micellar Polymer Flooding (MPF) represents an innovative approach within the realm of EOR. It leverages the combined benefits of both micellar flooding and polymer flooding, resulting in a synergistic strategy aimed at optimizing oil recovery from reservoirs. The MPF technique integrates a viscosifying agent, typically a water-soluble polymer, to enhance sweep efficiency significantly. By elevating the injected water viscosity, these water-soluble polymers effectively control mobility, isolate high-permeability regions within the reservoir, minimize oil bypass, and enhance the sweeping efficiency, particularly for less viscous crude oils [6, 7]. In certain specific conditions, the introduction of a surfactant into the polymer solution reduces capillary forces by reducing the IFT at the O/W interface. This alteration in IFT can lead to the re-mobilization of trapped oil, modify the surface's wettability, create emulsions, and ultimately improve displacement efficiency and oil production. This is particularly beneficial for recovering oil that is typically left unrecovered in conventional waterflooding methods [8]. In the absence of a polymer within the surfactant injection, the surfactant tends to channel and penetrate the oil bank, resulting in a highly ineffective reservoir sweep [9]. Moreover, the surfactant itself leads to an elevation in the relative permeability of water, a change that necessitates compensation by reducing water mobility through the addition of polymer. Additionally, incorporating polymer into both the surfactant injection and the subsequent drive slug serves to alleviate the impact of permeability variations and ultimately enhances the overall efficiency of the reservoir sweep. Consequently, including polymers within a surfactant slug is nearly indispensable to maintaining a favorable mobility balance [10]. This concern has been documented in prior research, laboratory experiments, and practical field tests, dating back to the initial investigations conducted by Marathon Oil Company in the 1960s [11]. Despite the extensive efforts invested by oil companies, and universities, throughout the 1970s and 1980s, which significantly expanded our understanding of chemical micellar/polymer flooding, ongoing research aimed at enhancing practical techniques persisted. This research encompassed several key areas: (1) simplifying the flooding process, (2) enhancing the efficiency of flooding chemicals, and (3) developing novel chemicals such as polymers, surfactants, and micellar/polymer flooding agents. In the forthcoming chapter, we will delve into the core principles, mechanisms, and practical applications of Micellar Polymer Flooding, shedding light on its potential, advantages, and limitations, with a particular focus on its capacity to alter wettability and potentially transform the oil and gas industry.

2. Principles of micellar polymer flooding (MPF)

MPF principles revolve around the integration of surfactants and polymers to enhance oil recovery from reservoirs. MPF combines the well-established mechanisms of micellar flooding and polymer flooding to create a synergistic approach. Surfactants are used to form micelles in the presence of oil and water, lowering the interfacial tension between these phases and solubilizing trapped oil. The addition of polymers increases the injected fluid viscosity, improving mobility control and sweep efficiency. Synergy between surfactants and polymers results in enhanced oil displacement and recovery. MPF's success is rooted in its ability to optimize the solubilization of oil, mobilize it efficiently, and control its movement through the reservoir, making it a promising method for maximizing oil production in various geological settings.

2.1 Polymer flooding

On the contrary, polymer flooding entails injecting water-soluble polymers into the reservoir. This process boosts the water viscosity being injected, which enhances sweep efficiency and controls mobility. The increased viscosity of the fluid being injected diminishes the ratio of water-oil mobility, resulting in improved oil displacement and heightened recovery. Polymers are used in processes aimed at altering permeability through selective plugging and as agents for controlling mobility [1]. The size of the polymer slug used during flooding typically falls within the range of 0.5–2.0 g/l. The volume of polymer solution injected can vary up to 50% of the pore volume, depending on the specific process design. Developing efficient polymer structures is of paramount importance in the context of oil and gas extraction. Polymer flooding represents the most widely utilized chemical method for EOR in both sandstone and carbonate reservoirs. To date, the literature references or reports more than 290 field projects involving polymers. Among these, studies of over 200 polymer floods have shown an average injection rate of 19–150 pounds per acre-foot and concentrations ranging from 50 to 3700 parts per million (ppm) [12]. Dabbous [13] Performed flood tests in diverse porous materials and demonstrated that introducing polymer before injection could enhance flooding efficiency. Additionally, the pre-injection of polymer did not impact the oil displacement properties of the micellar fluid, suggesting that it primarily reduced surfactant adsorption on the rock in the studied polymer-micellar system. Farouq-Ali and Thomas [14] reported average incremental oil recoveries of approximately 5%, suggesting the utilization of biopolymers like xanthan gum, either in powdered or gel form, as a recommended approach. Du and Guan [15] found that in low-permeability reservoirs, the injectivity of floods significantly decreased with the addition of polymer, leading to poor overall performance. El-hoshoudy et al. [16–26] performed polymer flooding experiments using synthetic polymers, biopolymers, and their modified composites incorporating vinyl monomers and nanoparticles like silica and other metal oxides. The outcomes indicated increased recovery factors in sandstone reservoirs. Additionally, they conducted simulation studies both on a computational basis and at the field scale.

2.2 Surfactant (Micellar) flooding

Micellar flooding is a recognized EOR technique that hinges on the application of surfactants to generate micelles within the reservoir. Micelles are colloidal structures created by surfactant molecules when water and oil are present [27, 28]. These micelles possess a hydrophobic core and a hydrophilic shell, making them adept at dissolving and mobilizing trapped oil within the reservoir. The surfactant molecules reduce IFT at the O/W interface, facilitating more efficient oil displacement [29, 30]. Surfactants are amphiphilic surface-active molecules, characterized by.

1. *Amphiphilic Nature:* Surfactants are inherently amphiphilic, possessing both hydrophilic (water-loving) and hydrophobic (water-repelling) components. So, they interact with aqueous and non-aqueous media, making them versatile in various applications.

2. *Hydrophilic Head and Hydrophobic Tail:* The hydrophilic head can be of different types (anionic, cationic, amphoteric, or nonionic), influencing how the surfactant interacts with different substances. The hydrophobic tail usually consists of long hydrocarbon chains.

3. *Adsorption at Interfaces:* Surfactants tend to adsorb at the interface between different phases, such as oil/water (O/W) interfaces.

4. *Critical Micelle Concentration (CMC):* CMC is the total concentration of surfactants above which micelles are formed, i.e., above the CMC surfactants are present in monomeric and micellar state, below CMC it can be approximated that surfactants are in monomeric state. At CMC, surfactants spontaneously form micelles. In micelles, the hydrophobic tails are in the center, shielded from water by the hydrophilic heads. This phenomenon is critical in reducing the free energy of the system.

5. *Interfacial Tension (IFT) Reduction:* By forming micelles and adsorbing at interfaces, surfactants reduce the interfacial tension between two phases. IFT reduction is crucial in processes like emulsification, where surfactants help in mixing oil and water, and in detergency, where they help remove dirt and grease from surfaces. The understanding of these properties and behaviors of surfactants is fundamental in fields such as chemistry, petroleum engineering, and enhanced oil recovery (EOR) where they are used in a wide range of applications from detergents to drug delivery systems [10, 31]. Surfactant flooding involves injecting a dilute solution of surfactant into underground petroleum reservoirs. This process leads to IFT reduction at the O/W interface, generally improving the movement of residual oil trapped within the pores of the reservoir rocks [32]. Key characteristics of a favorable surfactant for EOR applications in reservoirs involve the following aspects.

   a. *Low Adsorption on Reservoir Rocks:* to maintain its efficacy in the reservoir. High adsorption can lead to loss of surfactant and reduced efficiency in reducing IFT.

   b. *Long-term Stability at Reservoir Conditions:* Surfactants must be stable over large temperatures scale and under the specific conditions of the reservoir, including brine salinity and hardness. The stability ensures that the surfactant retains its effectiveness throughout the EOR process.

   c. *Compatibility with Reservoir Fluids:* The surfactant must be compatible with the reservoir fluids, particularly in its tolerance to divalent (Ca^{2+}) and (Mg^{2+}) cations. These ions are commonly present in reservoirs and can significantly impact the performance of surfactants.

   d. *IFT Reduction:* One of the primary roles of surfactants in EOR is to decrease the IFT between oil and water to extremely low levels, typically around 0.001 mN/m. This action significantly raises the capillary number (Nc) by several orders of magnitude, reducing capillary forces and altering the oil contact angle. As a result, the wettability of the reservoir is modified, leading to improved mobilization of trapped oil and ultimately enhancing the overall oil recovery process [33].

   e. *Correlation with Capillary Number (N_c):* The capillary number (N_c) is a dimensionless number that quantifies the relative effect of viscous forces over capillary forces in a porous medium, like reservoir rocks. IFT correlated with capillary number (Nc) through Eq. (1) [34].

$$Nc = \frac{\mu v}{\sigma} \tag{1}$$

A higher capillary number indicates that viscous forces dominate, which is desirable in EOR as it implies easier displacement of oil by water. Surfactants play a crucial role in this by reducing the IFT (σ), thus increasing the capillary number and enhancing oil displacement.

a. *Altering Rock Wettability:* Changing the matrix wettability to favor more water-wet conditions can lead to an increase in the rate at which brine is imbibed into the rock. In addition, Tailor the attributes of polymer systems to suit various purposes, encompassing particle scattering, emulsion steadying, foam formation, and altering reservoir wetting. Generally, an ideal surfactant for reservoir applications should minimize adsorption to reservoir rocks, be stable under reservoir conditions, compatible with reservoir fluids, and effectively reduce IFT to increase the capillary number for efficient oil recovery. Formulating middle-phase microemulsions using a suitable combination of surfactant and co-surfactant or solvent can result in achieving extremely low interfacial tension (IFT) values [34]. Anionic surfactants are widely used in chemical-EOR methods due to their lower tendency to adsorb on negatively charged sandstone rock surfaces [10]. Conversely, the implementation of surfactant flooding in enhanced oil recovery procedures is often regarded as financially impractical and continues to pose difficulties, particularly in environments characterized by elevated salinity and high temperatures, primarily because of the following limitations.

1. The retention of chemicals within porous media through adsorption, as observed by Somasundaran and Zhang [35], plays a significant role in determining the economic viability of oil recovery or remediation processes.

2. Surfactant aggregation: The formation of surfactant aggregates is associated with a limited ability to withstand the presence of divalent ions, high salinity, and elevated temperatures exceeding 90°C.

3. Following the findings of Taugbøl et al. [36] and Berger and Lee [37], high-performance surfactants are effective at significantly reducing the interfacial tension between oil and water. However, they do not promote capillary-driven imbibition when employed in water flooding processes.

2.2.1 Types of Winsor's Type phase behavior

In a water-oil system, a portion of the surfactant monomers will tend to cluster or aggregate at the interface between the oil and water phases. This aggregation of surfactant molecules serves to decrease the interfacial tension between the two immiscible phases. Interestingly, it's worth noting that only a relatively small quantity of surfactant is required to reach a point where the interface becomes saturated with surfactant molecules. Winsor's Type phase behavior refers to a classification system developed by Winsor to categorize the equilibrium phases formed when oil, water, and surfactants are mixed. This classification is especially relevant in the study of microemulsions and their applications, such as EOR processes. Winsor identified three main types of phase behavior [8].

1. *Winsor Type I*: occurs at low electrolyte concentrations when the surfactant has a higher solubility in water than in oil. This leads to the creation of an oil-in-water

(O/W) microemulsion, where the oil phase is dispersed as small droplets within a continuous water phase. In this type of microemulsion, the microemulsion structure coexists with an excess oil phase, indicating the presence of both water and oil phases in equilibrium. O/W microemulsions are commonly encountered in various applications, including emulsification processes and the formulation of certain types of products like creams and lotions in the cosmetics industry.

2. *Winsor Type II:* In the scenario described, when the surfactant concentration exceeds the CMC and the electrolyte concentration is high, the micelles exhibit a stronger attraction to the oil phase. In this situation, the micelles can incorporate or solubilize brine into the oleic (oil) phase, leading to the formation of a water-in-oil (W/O) microemulsion, in which, the micellar structures coexist with an excess water phase, indicating the presence of both water and oil phases in equilibrium. This behavior can have important implications in processes such as emulsification and microemulsion-based enhanced oil recovery, where the ability to create stable W/O microemulsions can be advantageous for specific applications.

3. *Winsor Type III*: occurs by the creation of a central microemulsion phase, which exists alongside surplus water and oil phases. This phenomenon takes place when the surfactant can distribute itself equally between oil and water, resulting in a bicontinuous microemulsion. Comprehending the phase behavior as described by Winsor is essential in developing efficient surfactant systems for various uses, such as EOR. In EOR, the appropriate microemulsion can greatly improve oil recovery effectiveness by aiding in the release of oil trapped within reservoirs. The phase containing the micelles, which also dissolves the other phases, is termed the microemulsion phase and is known for its thermodynamic stability. Phases that are not part of the microemulsion are referred to as excess phases. These are categorized as pure when they lack micelles, or impure if they contain some micelles. In a specific electrolyte concentration range, a bi-continuous micellar structure emerges, leading to the formation of a middle microemulsion phase alongside the excess oil and water phases, a condition identified as Winsor's Type III phase environment.

4. *Winsor Type IV:* In situations with elevated surfactant concentrations, a solitary microemulsion phase emerges, sometimes termed Winsor's Type IV phase configuration by certain researchers. This phase is distinguished by the creation of a cohesive microemulsion, which does not coexist with separate, excess water or oil phases, unlike the preceding types. This unique phase behavior occurs when the concentration of the surfactant is sufficiently high to completely blend the oil and water components into a single, homogenous microemulsion [8]. A depiction of the microemulsion systems for Types I, II, and III is shown in **Figure 1**.

Several instances of technically successful surfactant field projects, conducted at both laboratory and field scales, have been recognized in the scientific literature [38]. In their study, Boneau and Clampitt [39] carried out core flood experiments in sandstones that exhibited similar characteristics in terms of porosity, permeability, and pore structure. They used a surfactant and observed that in the oil-wet sandstones, tertiary oil recoveries varied between 55% and 65%, while in the water-wet sandstones, they ranged from 90–95%. In Lawson's research Lawson et al. [40], an

DOI: http://dx.doi.org/10.5772/intechopen.114288

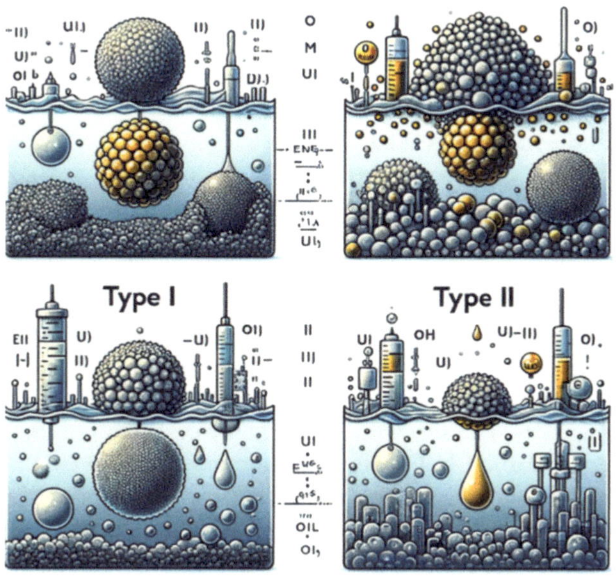

Figure 1.
Winsor's microemulsion systems [8].

investigation was carried out regarding the adsorption of both nonionic and cationic surfactants on sandstones and carbonates. The adsorption of carbonates was lower compared to sandstone. Regarding cationic surfactants, the adsorption isotherms followed Langmuir's behavior, and it was observed that multivalent cations led to an increase in adsorption. As reported by Chou and Bae [41], solutions of oligomeric surfactants, when used at concentrations approximately one-tenth of that of active surfactants, exhibit several distinguishing characteristics. These include greater solubilization capacity, reduced IFT, increased optimal salinity, enhanced viscosity and salt tolerance, and decreased adsorption loss. In a study conducted by Bae [42], a micellar flooding project was implemented in Chevron's Glenn Pool Field located in Oklahoma. The project resulted in the recovery of one-third of the remaining oil saturation from shallow, low-permeability sandstone reservoirs. Wu [43] researched optimizing surfactant flooding. The primary focus of the study was onshore sandstone reservoirs, with specific emphasis on optimizing chemical concentrations, and slug sizes and addressing issues related to adsorption. Berger and Lee [44] developed a novel approach in which they synthesized anionic surfactants by simultaneously alkylating and sulfonating aromatic compounds. This innovative family of anionic surfactants exhibits the ability to achieve ultra-low interfacial tensions even at low concentration levels. Additionally, they possess a high tolerance to salt levels, and their use can help minimize issues related to emulsions and corrosion. In their study, Li et al. [45] observed that surfactants featuring branched hydrocarbon chains led to a significant reduction in both advancing and receding contact angles, approximately by 300. This effect highlights an enhanced efficiency in promoting water-wetting. The researchers attributed this phenomenon to two main factors: the increased structural rigidity of the surfactant's branched hydrophobic tail and a broader area of coverage provided by these molecules. Bhui et al. [32] utilize fluorescence spectroscopy as a method to investigate the behavior of micellar flooding in Enhanced Oil Recovery (EOR) processes. Cao et al. [46] studied the spontaneous imbibition of aqueous surfactant solutions into preferentially oil-wet carbonate cores. Zhao et al. [34] employ naphthenic aryl sulfonate for in-situ emulsification as a strategy to enhance oil recovery, leveraging its blocking and entrainment effects.

2.3 Micellar/polymer flooding (MPF)

In surfactant flooding, the formation of emulsions, whether oil in water (O/W) or water in oil (W/O), is due to low IFT, which enhances displacement efficiency. These emulsion droplets merge to create an oil ganglia in the surfactant front, making the oil bank mobile. In contrast, the primary mechanism in polymer flooding is to enhance sweep efficiency. Therefore, in micellar-polymer (MPF) flooding, the key mechanism is the combined effect of these two processes – the mobilization of oil through surfactant-induced emulsion formation and the improved sweep efficiency provided by polymer flooding [10]. The screening criteria for micellar/polymer flooding concluded as indicated in **Table 1** [10]. Micellar Polymer Flooding (MPF) has exhibited promising outcomes in both laboratory experiments and field applications. Numerous case studies have highlighted its effectiveness across diverse reservoir types, such as sandstone and carbonate formations. Often, MPF has resulted in considerable incremental oil recovery, establishing it as a valuable component in the array of Enhanced Oil Recovery (EOR) techniques.

Few micellar/polymer flooding projects are conducted, as they include low-tension waterflooding, and sequential micellar/polymer flooding [10]. **Figure 2** provides an overview of the research papers and articles related to micellar polymer flooding as documented in the Scopus database based on their relevant keywords. In the mid-1960s, Mobil initiated a low-tension field test in a water-flooded section of the Loma Novia field, located in Duval County, Texas, as noted by Sheng in 2013. Additionally, the Cities Service Oil Company, in collaboration with the Energy Research and Development Administration, carried out a field demonstration project of MPF flooding in the El Dorado field, situated in Butler County, Kansas [10]. Osterloh and Jante [47] stated that surfactant adsorption is reduced in the presence of polymer, making Micellar Polymer Flooding (MPF) highly advantageous and optimal for Enhanced Oil Recovery (EOR) operations. Schramm [48] introduced the concept of "Low Tension Polymer Flood", a technique that combines polymer flooding with micellar flooding. In this process, the polymer is designed to move ahead of the surfactant, leading to reduced water mobility at the front and enhanced surfactant activity. The use of ethoxylated sulfonate and scleroglucan (a type of polymer) as LTPF chemicals has shown promising results, particularly in conditions of high temperature and salinity. These chemicals demonstrated effective recovery characteristics, achieving 66% recovery,

Screening parameter	Value
CL^- ions Concentration	< 20,000 ppm
Divalent cations (Ca^{+2}, Mg^{+2}) concentration	< 500 ppm
Reservoir Temperature, °C	<90–95°C
Oil Viscosity, cP	<35 cP
Absolute permeability	> 10 mD

Table 1.
The screening standards for micellar/polymer flooding (MPF).

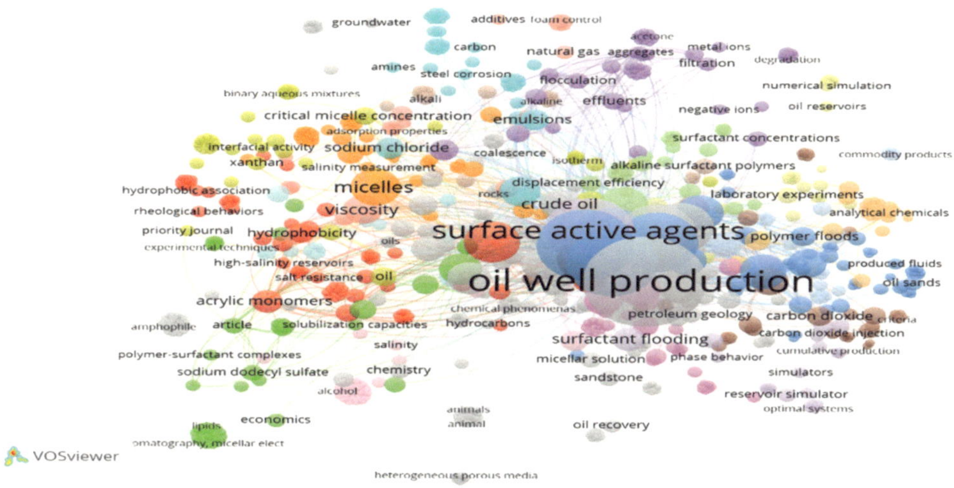

Figure 2.
An overview of the published literature as documented in the Scopus database.

and exhibited low surfactant retention, with only 0.18 mg/g of rock in core samples up to 2.9 m. One notable benefit of the simultaneous injection of surfactant and polymer, as observed in LTPF, is likely the improvement in sweep efficiency.

Al-Ghailani et al. [1] A novel combination for EOR involves the use of surfactin, a biosurfactant, and schizophyllan, a biopolymer, assisted by sodium carbonate (Na_2CO_3) as an alkali agent. Surfactin, known for its surface-active properties, helps reduce O/W-IFT, enhancing oil mobilization. Schizophyllan, as a biopolymer, increases the viscosity of the flooding solution, improving sweep efficiency in the reservoir. The addition of sodium carbonate, an alkali, further aids the process by altering the rock wettability and potentially generating in-situ soap, which can improve oil recovery. This combination presents a unique and environmentally friendly approach to EOR, utilizing naturally derived materials. de Castro Dantas et al. [3] Examined the impact of injecting a high water content microemulsion formulation on rock-fluid interactions and the wettability of Botucatu sandstone rock. Fernandes et al. [8] developed fully implicit (FI) approaches that are key for accurately detecting the phase behavior of polymer/microemulsions under simulated reservoir conditions. These approaches offer a detailed and comprehensive overview of the developed numerical models, which are specifically tailored for simulating the phase behavior of polymer/micellar systems. For those interested in a deeper understanding of these methods, further reading on the subject would provide valuable insights into the intricacies of these models and how they effectively simulate complex fluid interactions and phase behaviors in reservoir environments.

Gao et al. [49] reported a blend of hydrolyzed polyacrylamide and sodium petroleum sulfonate has been utilized not only for enhancing oil displacement but also as a corrosion inhibitor. This dual-function approach leverages the unique properties of both compounds: hydrolyzed polyacrylamide, known for its effectiveness in improving oil recovery through boosting the viscosity of the flooding solution, and sodium petroleum sulfonate, recognized for its surfactant properties and ability to reduce corrosion in the oil recovery infrastructure. This combination thereby serves a dual purpose in Enhanced Oil Recovery (EOR) operations, optimizing oil displacement while simultaneously protecting equipment from corrosion.

3. Synergies of micellar and polymer flooding

The synergies between micellar and polymer flooding present a significant advancement in EOR techniques. Micellar flooding, involving the injection of surfactants, effectively reduces O/W-IFT, facilitating the mobilization of trapped oil. However, its efficiency can be limited by poor sweep efficiency due to the surfactant's tendency to flow through high permeability paths. This is where polymer flooding complements the process. By injecting polymers, the viscosity of the displacing water is improved, improving sweep efficiency, and ensuring a more uniform displacement front. This combination allows for the extensive mobilization of oil by micellar flooding while ensuring a more effective sweep of the reservoir by polymer flooding. The result is an enhanced overall oil recovery, making the synergy of these two methods a powerful tool in EOR operations. The synergy between surfactants and polymers can be summarized as follows:

3.1 Enhanced solubilization

The surfactants in the micellar phase solubilize oil and reduce interfacial tension, making it easier to mobilize and displace trapped oil. The addition of polymers increases micellar phase viscosity, further improving its ability to carry and transport oil droplets.

3.2 Mobility control

Polymers, by increasing the injected fluid viscosity, create a favorable mobility contrast between the reservoir fluids and injected fluid. This mobility control helps push the micellar phase deeper into the reservoir, ensuring better sweep efficiency and minimizing bypassed oil zones.

3.3 Shear-thinning behavior

Many polymer solutions exhibit pseudoplastic behavior, which means their viscosity is reduced under shear forces. This property allows for easier injection of the micellar polymer solution through the reservoir, reducing energy consumption and wellhead pressures during the injection process.

4. Design of micellar/polymer injection scenario

Due to initial challenges with injectivity and ongoing issues in this domain, significant attention was given to a water-quality monitoring program. This program was established to rigorously track the quality of injected fluids, with testing scheduled daily. The objective was to ensure that the characteristics of the water being injected met the necessary standards and conditions for optimal performance. Regular monitoring allowed for the early detection and mitigation of potential problems that could impact the effectiveness of the injection process, such as the presence of contaminants or deviations in desired chemical composition. This proactive approach was crucial in maintaining the efficiency and success of the injection operations [10]. Designing a micellar/polymer injection scenario requires a careful balance of various factors to optimize EOR. Firstly, surfactant selection for the micellar solution is critical. These must be capable of forming stable microemulsions with reservoir oil,

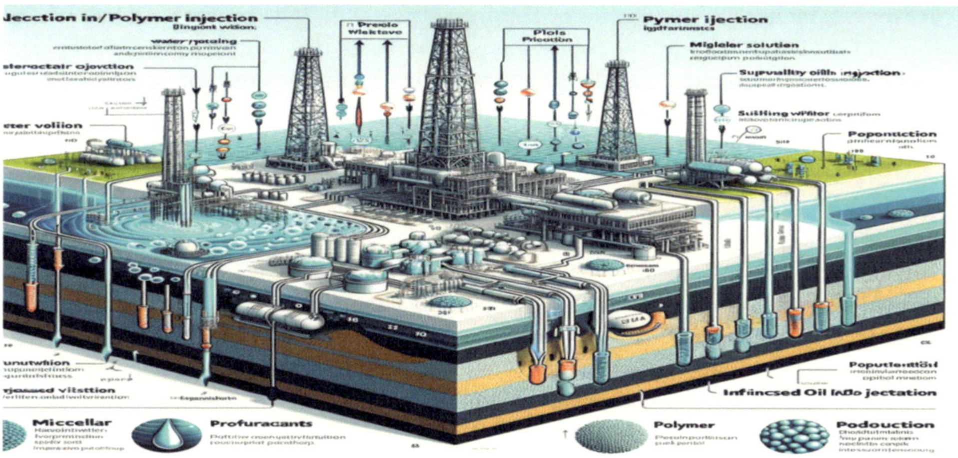

Figure 3.
Injection pattern for micellar/polymer injection.

reducing interfacial tension effectively. The concentration of surfactants needs to be tailored to reservoir conditions, ensuring optimal emulsification without excessive adsorption on reservoir rocks. Next, the choice of polymer and its concentration is crucial for increasing the injected fluid viscosity, thus enhancing sweep efficiency. This needs to be compatible with the surfactant system and resilient to reservoir conditions such as temperature and salinity. Additionally, the injection strategy as shown in **Figure 3** including the sequence and ratio of micellar and polymer solutions- must be meticulously planned. This involves deciding whether to inject them simultaneously or sequentially and determining the appropriate volumes and injection rates. The scenario design also considers reservoir characteristics like permeability, porosity, and existing water saturation, as well as the economic aspects to ensure a cost-effective and efficient recovery process. Overall, the design of a micellar/polymer injection scenario is a complex task that combines chemical engineering, reservoir engineering, and economic analysis to achieve the best possible EOR results.

5. Reservoir wettability

EOR through rock's wettability alteration is a well-established technique in the industry. Accurate determination of a reservoir's original wettability is crucial; incorrect assessment can result in suboptimal decisions regarding EOR field applications that utilize surfactants. Therefore, selecting the appropriate surfactant, based on the initial reservoir wettability, is essential to maximize the effectiveness of the EOR process. The surfactant chosen must align with the reservoir's characteristics to ensure it effectively modifies the wettability, thereby enhancing oil recovery by facilitating the release and mobilization of oil trapped in the reservoir. Wettability Change stimulates capillary pressure changes, and residual oil saturation [50, 51]. Jadhunandan and Morrow [52] explored the connection between wettability and oil recovery through water flooding in various Berea core samples. The research revealed that as the wettability shifted from predominantly water-wet toward oil-wet, the efficiency of oil recovery by water flooding initially rose and subsequently declined. Johannesen and Graue [53] introduced a methodical approach to examine the recovery of oil in chalk, focusing on how it is influenced by wettability. Zhao et al. [54] projected a hypothetical model for wettability change following primary drainage, suggesting that regions of the pore space that come into direct contact with oil undergo a change in their oil/water contact angle in preparation for water flooding. According to this model, the areas of the pore space that are initially in contact with oil alter their wettability, potentially becoming more oil-wet or less water-wet. This change in wettability is crucial for enhancing oil recovery during subsequent water flooding. However, the regions of the pore space that remain filled with water retain their original water-wet characteristics. This differentiation in wettability within the reservoir can significantly influence the efficiency of water flooding as an EOR method, as it affects the distribution and flow of injected water and its ability to displace oil. Noruzi et al. [55] offer an in-depth overview of the current state of wettability modification techniques in sandstone and carbonate reservoirs:

6. Evaluation of micellar characterization

Various techniques and analyses were conducted to inspect the micelles criteria. The (IFT) was assessed using a spinning drop tensiometer. Contact angles were measured under both static and dynamic conditions using tensiometers. The size of the micelles was determined using a Zetasizer Nano at a temperature of 40°C. The micelles solubilization capacity is measured by light absorbance with a UV-Vis spectrophotometer. Rheological studies of the microemulsion were conducted using a Haake Maars rheometer (Thermo Scientific) [3, 34].

7. Flooding setup

The flooding setup for micellar polymer flooding is a critical component of EOR operations. It involves a meticulously designed system for injecting a combination of micellar and polymer solutions into the reservoir [20]. The setup typically includes injection wells strategically placed within the reservoir to introduce the fluids into the subsurface. Micellar solutions, containing surfactants, are injected to alter the wettability of the reservoir rock and mobilize trapped oil. Monitoring and control systems are integrated to manage injection rates, pressures, and fluid compositions, optimizing the injection process. Extensive subsurface data, reservoir characteristics, and fluid properties are considered during setup design to tailor the EOR approach to the specific reservoir conditions. The flooding setup for micellar polymer flooding is a sophisticated engineering solution aimed at maximizing oil recovery from challenging reservoirs. Micellar polymer flooding is an enhanced oil recovery (EOR) technique used in the oil industry to improve oil recovery from reservoirs. It involves the injection of surfactants and polymers to increase the displacing efficiency and recover more oil. On the lab scale, the micellar polymer flooding set-up comprises the following components:

1. *Core Samples:* prepare core samples from a representative rock formation similar to the reservoir you want to study. These core samples should be cylindrical and should have the same properties as the reservoir rock.

2. *Core Holder*: A core holder is used to hold the core sample in place during the flooding experiment. It allows you to control the flow of fluids through the core.

3. *Injection Pump*: used to inject the micellar polymer solution into the core sample. Ensure the pump can deliver a constant and controlled flow rate.

4. *Injected Fluids*: comprise all fluids used in the injection scenario, as follows.

    - *Micellar Solution*: Prepare a micellar solution by mixing surfactants, typically anionic or nonionic, with water. The surfactants reduce interfacial tension, allowing the oil to be displaced more easily.

    - *Polymer Solution*: Prepare a polymer solution by dissolving a suitable polymer in water. The polymer improves sweeping efficiency by increasing the injected fluid viscosity.

5. *Pressure and Flow Measurement Instruments:* Install pressure gauges and flow meters at various points in the system to monitor pressure drop and flow rates during the flooding experiment.

6. *Porous Plate and Filter:* used to ensure that the injected solutions are properly mixed before they enter the core sample. This helps in the uniform injection of the micellar polymer solution.

7. *Data Acquisition System*: Set up a data acquisition system to record pressure, flow rates, and other relevant data during the experiment.

8. *Injection and Production Lines:* Connect injection and production lines to the core holder to facilitate fluid injection and oil recovery.

The experimental procedure involves injecting the micellar polymer solution into the core sample and monitoring the pressure and flow rates over time. The goal is to observe the displacement of oil from the core sample and measure the incremental oil recovery achieved through the micellar polymer flooding process. It's crucial to maintain controlled experimental conditions, including temperature, pressure, and flow rates, to simulate the reservoir conditions as accurately as possible. Data collected during the lab-scale experiment can help in optimizing the micellar polymer flooding technique for field-scale applications.

8. Future prospects

1. The creation of novel models and fully implicit methods for surfactant-polymer flooding, encompassing all significant phase behavior categories.

2. Application Expansion: MPF may find broader applications in various types of reservoirs, including unconventional resources like shale oil and gas, where maximizing recovery is critical.

3. Environmental Considerations: As sustainability becomes increasingly important, MPF's ability to improve oil recovery with fewer wells and reduced environmental impact will make it an attractive option for environmentally conscious operators.

4. Integration with Digital Technologies: The integration of digital technologies, such as artificial intelligence and machine learning, can optimize MPF processes by predicting reservoir behavior and automating decision-making.

5. Pilot Projects: More pilot projects and field trials will provide valuable data and real-world insights, further demonstrating MPF's effectiveness.

In summary, the future of micellar polymer flooding holds significant potential for the oil and gas industry. Advances in technology, environmental awareness, and the need for sustainable resource management are likely to drive the continued development and adoption of MPF as a valuable tool for maximizing oil recovery from reservoirs.

9. Conclusion

Micellar Polymer Flooding (MPF) indeed represents a pioneering and promising approach to EOR within the gas and oil industry. By effectively combining the benefits of both micellar and polymer flooding techniques, MPF offers a compelling solution for improving oil recovery from reservoirs. This approach addresses several key challenges in the industry and has the potential to revolutionize oil production in several ways:

1. Improved Recovery Efficiency: MPF can significantly enhance the displacement of trapped oil within reservoirs. The combination of surfactants to reduce IFT and polymers to increase fluid viscosity leads to improved sweep efficiency and oil recovery rates.

2. Sustainable Resource Utilization: As the global energy demand continues to rise, it is crucial to optimize oil production from existing reservoirs while minimizing environmental impacts. MPF provides a more efficient method for accessing and recovering oil, contributing to sustainable energy production.

3. Economic Viability: Enhanced oil recovery techniques like MPF can increase the profitability of oil production operations. Higher oil recovery rates translate into increased revenue and extend the economic life of reservoirs, making it an attractive option for operators.

4. Technological Advancements: As technology continues to advance, the understanding and application of MPF are expected to improve further. Ongoing research, experimentation, and field trials will likely refine the methodology and lead to even greater success in maximizing oil recovery.

5. Environmental Benefits: By recovering more oil with fewer wells and reducing the need for excessive drilling, MPF can help mitigate environmental impacts associated with oil and gas extraction, including habitat disruption and greenhouse gas emissions.

Overall, as the oil and gas industry seeks more efficient and sustainable methods of resource extraction, Micellar Polymer Flooding is poised to play an increasingly vital role. With further development, field trials, and the integration of emerging technologies, MPF has the potential to contribute significantly to sustainable energy production, resource optimization, and the ongoing evolution of the industry. As such, it represents a promising path toward a more efficient and environmentally responsible future for oil recovery.

Abbreviation

EOR	enhanced oil recovery
MPF	Micellar polymer flooding
IFT	interfacial tension
CMC	critical Micelle concentration
Nc	capillary number, dimensionless quantity
μ	viscosity, Cp
v	the Darcy velocity of the driving fluid
σ	the oil-water IFT

Author details

Abdelaziz El-Hoshoudy[1,2]

1 Production Department, PVT-Lab, Egyptian Petroleum Research Institute, Cairo, Egypt

2 PVT-Service Center, Egyptian Petroleum Research Institute, Cairo, Egypt

*Address all correspondence to: abdelaziz.nasr@epri.sci.eg; azizchemist@yahoo.com

References

[1] Al-Ghailani T, Al-Wahaibi YM, Joshi SJ, Al-Bahry SN, Elshafie AE, Al-Bemani AS. Application of a new bio-ASP for enhancement of oil recovery: Mechanism study and core displacement test. Fuel. 2021;**287**:119432

[2] Salem KG, Tantawy MA, Gawish AA, Gomaa S, El-hoshoudy A. Nanoparticles assisted polymer flooding: Comprehensive assessment and empirical correlation. Geoenergy Science and Engineering. 2023;**226**:211753

[3] de Castro Dantas TN, Viana FF, de Souza TTC, Neto AAD, Aum PTP. Study of single-phase polymer-alkaline-microemulsion flooding for enhancing oil recovery in sandstone reservoirs. Fuel. 2021;**302**:121176

[4] Khattab H, Gawish AA, Hamdy A, Gomaa S, El-hoshoudy A. Assessment of a novel xanthan gum-based composite for oil recovery improvement at reservoir conditions; assisted with simulation and economic studies. Journal of Polymers and the Environment. 2024:1-29

[5] Schramm LL, Isaacs EE. Foams in enhancing petroleum recovery. In: Foam Engineering: Fundamentals and Applications. Wiley Online Library; 2012. pp. 283-305

[6] Zhang X, Li B, Pan F, Su X, Feng Y. Enhancing oil recovery from low-permeability reservoirs with a Thermoviscosifying water-soluble polymer. Molecules. 2021;**26**(24):7468

[7] El-Hoshoudy A, Desouky S, Betiha M, Alsabagh A. Use of 1-vinyl imidazole based surfmers for preparation of polyacrylamide–SiO_2 nanocomposite through AZA-Michael addition copolymerization reaction for rock wettability alteration. Fuel. 2016;**170**:161-175

[8] Fernandes BRB, Sepehrnoori K, Delshad M, Marcondes F. New fully implicit formulations for the multicomponent surfactant-polymer flooding reservoir simulation. Applied Mathematical Modelling. 2022;**105**:751-799

[9] El-Hoshoudy A, Desouky S, Al-Sabagh A, Betiha M, MY E-k, Mahmoud S. Evaluation of solution and rheological properties for hydrophobically associated polyacrylamide copolymer as a promised enhanced oil recovery candidate. Egyptian Journal of Petroleum. 2017;**26**(3):779-785

[10] Sheng JJ. Surfactant-polymer Flooding: Enhanced Oil Recovery Field Case Studies. Elsevier; 2013. pp. 117-142

[11] Austad T, Milter J, Schramm L. Surfactant flooding in enhanced oil recovery. In: Surfactants: Fundamentals and Applications in the Petroleum Industry. Cambridge University Press; 2000. pp. 203-249

[12] El-hoshoudy AN, Desouky SM, Betiha MH, Alsabagh AM. Hydrophobic polymers flooding. In: Application and Characterization of Surfactants. London, UK: InTechopen; 2017. pp. 75-95

[13] Dabbous MK. Displacement of polymers in waterflooded porous media and its effects on a subsequent micellar flood. Society of Petroleum Engineers Journal. 1977;**17**(05):358-368

[14] Farouq Ali S, Thomas S. A realistic look at enhanced oil recovery. Scientia Iranica. 1994;**1**(3):220-230

[15] Du Y, Guan L. Field-scale polymer flooding: Lessons learnt and experiences gained during past 40 years. In: SPE International Oil Conference and Exhibition in Mexico. SPE; 2004. SPE-91787-MS

[16] El-Hoshoudy AN, Elsaeedb SM. Experimental and Monte Carlo simulation of palmitate-guar gum derivative as a novel flooding agent in the underground reservoir. Journal of Molecular Liquids. 2020;**302**:112502

[17] El-Hoshoudy AN, Desouky SM, Ramzi M, Attia AM. Experimental, modeling and simulation investigations of a novel surfmer-co-poly acrylates crosslinked hydrogels for water shut-off and improved oil recovery. Journal of Molecular Liquids. 2019;**277**:142-156

[18] El-Hoshoudy AN, Fathy M, Abdelraheem OH, Gomaa S, Desouky SM. Enhanced oil recovery using polyacrylates/ACTF crosslinked composite: Preparation, characterization and coreflood investigation. Journal of Petroleum Science and Engineering. 2019;**181**:106236

[19] Abou-alfitooh SA, El-Hosiny F, Ramzi M, Mansour E, Elnaggar OM, El-hoshoudy A. Chemical modification of guar by different synthetic vinyl monomers for enhancing oil recovery under severe sandstone reservoir conditions. Egyptian Journal of Petroleum. 2021;**30**(3):35-43

[20] El-hoshoudy A. Experimental and theoretical investigation for synthetic polymers, biopolymers and polymeric nanocomposites application in enhanced oil recovery operations. Arabian Journal for Science and Engineering. 2022;**47**(9):10887-10915

[21] El-hoshoudy A, Mansour E, Desouky S. Experimental, computational and simulation oversight of silica-co-poly acrylates composite prepared by surfactant-stabilized emulsion for polymer flooding in unconsolidated sandstone reservoirs. Journal of Molecular Liquids. 2020;**308**:113082

[22] El-Hoshoudy AN. Experimental and theoretical investigation of glycol-based hydrogels through waterflooding processes in oil reservoirs using molecular dynamics and dissipative particle dynamics simulation. ACS Omega. 2021;**6**(45):30224-30240

[23] Soliman AA, El-Hoshoudy AN, Attia AM. Assessment of xanthan gum and xanthan-g-silica derivatives as chemical flooding agents and rock wettability modifiers. Oil & Gas Science and Technology–Revue d'IFP Energies nouvelles. 2020;**75**:12

[24] Abou-Alfitooh SA, El-hoshoudy A. Eco-friendly modified biopolymers for enhancing oil production: A review. Journal of Polymers and the Environment. 2023:1-27

[25] Gomaa S, Salem KG, El-hoshoudy A. Recovery of heavy oil and extra heavy oil; current status, new trends, and enhancement techniques. Petroleum. 2023 [In press, corrected proof]

[26] El-hoshoudy A, Hosny R, Fathy M, Abdelraheem O, Gomaa S, Desouky S. Enhanced oil recovery using polyacrylates/ACTF crosslinked composite: Preparation, characterization and coreflood investigation. Journal of Petroleum Science and Engineering. 2019;**181**:106236

[27] Afolabi F, Mahmood SM, Yekeen N, Akbari S, Sharifigaliuk H. Polymeric surfactants for enhanced oil recovery: A review of recent progress. Journal of Petroleum Science and Engineering. 2022;**208**:109358

[28] Liu Z, Zhao G, Brewer M, Lv Q, Sudhölter EJ. Comprehensive review on surfactant adsorption on mineral surfaces in chemical enhanced oil recovery. Advances in Colloid and Interface Science. 2021;**294**:102467

[29] Sagir M, Mushtaq M, Tahir MS, Tahir MB, Shaik AR. Surfactants for Enhanced Oil Recovery Applications. Springer; 2020

[30] Isaac OT, Pu H, Oni BA, Samson FA. Surfactants employed in conventional and unconventional reservoirs for enhanced oil recovery—A review. Energy Reports. 2022;**8**:2806-2830

[31] Chowdhury S, Shrivastava S, Kakati A, Sangwai JS. Comprehensive review on the role of surfactants in the chemical enhanced oil recovery process. Industrial & Engineering Chemistry Research. 2022;**61**(1):21-64

[32] Bhui UK, Sanyal S, Saha R, Rakshit S, Pal SK. Steady-state and time-resolved fluorescence spectroscopic study of petroleum crudes in aqueous-surfactant solutions: Its implications for enhanced oil recovery (EOR) during surfactant flooding. Fuel. 2018;**234**:1081-1088

[33] Khandoozi S, Sharifi A, Riazi M. Enhanced oil recovery using surfactants. In: Chemical Methods. 2022. pp. 95-139

[34] Zhao X-Z, Liao G-Z, Gong L-Y, Luan H-X, Chen Q-S, Liu W-D, et al. New insights into the mechanism of surfactant enhanced oil recovery: Micellar solubilization and in-situ emulsification. Petroleum Science. 2022;**19**(2):870-881

[35] Somasundaran P, Zhang L. Adsorption of surfactants on minerals for wettability control in improved oil recovery processes. Journal of Petroleum Science and Engineering. 2006;**52**(1-4):198-212

[36] Taugbøl K, Van Ly T, Austad T. Chemical flooding of oil reservoirs 3. Dissociative surfactant-polymer interaction with a positive effect on oil recovery. Colloids and Surfaces A: Physicochemical and Engineering Aspects. 1995;**103**(1-2):83-90

[37] Berger P, Lee C. Ultra-low concentration surfactants for sandstone and limestone floods. In: SPE Improved Oil Recovery Conference? Muscat, Oman: SPE; 2002. SPE-75186-MS

[38] Barnes JR, van Batenburg DW, Faber M, Van Rijn C, Geib S, van Kuijk SR, et al. Quality assurance and quality control of surfactants for field-scale enhanced-oil-recovery pilot projects. SPE Reservoir Evaluation & Engineering. 2019;**22**(02):565-576

[39] Boneau D, Clampitt RL. A surfactant system for the oil-wet sandstone of the North Burbank Unit. Journal of Petroleum Technology. 1977;**29**(05):501-506

[40] Lawson EE, Brown ER, Torday JS, Madansky DL, Taeusch HW Jr. The effect of epinephrine on tracheal fluid flow and surfactant efflux in fetal sheep. American Review of Respiratory Disease. 1978;**118**(6):1023-1026

[41] Chou S, Bae J. Phase-behavior correlation for high-salinity surfactant formulations. SPE Reservoir Engineering. 1988;**3**(03):778-790

[42] Bae J. Glenn pool surfactant-flood expansion project: A technical summary. SPE Reservoir Engineering. 1995;**10**(02):123-128

[43] Wu W-j. Optimum Design of Field-Scale Chemical Flooding Using Reservoir Simulation. The University of Texas at Austin; 1996

[44] Berger PD, Lee CH. New anionic alkylaryl surfactants based on olefin sulfonic acids. Journal of Surfactants and Detergents. 2002;**5**(1):39-43

[45] Li P, Yang C, Cui Z, Song B, Jiang J, Wang Z. A new type of sulfobetaine surfactant with double alkyl polyoxyethylene ether chains for enhanced oil recovery. Journal of Surfactants and Detergents. 2016;**19**:967-977

[46] Cao B, Lu X, Xie K, Ding H, Xiao Z, Cao W, et al. The pore-scale mechanisms of surfactant-assisted spontaneous and forced imbibition in water-wet tight oil reservoirs. Journal of Petroleum Science and Engineering. 2022;**213**:110371

[47] Osterloh W, Jante M Jr. Effects of gas and liquid velocity on steady-state foam flow at high temperature. In: SPE Improved Oil Recovery Conference? Muscat, Oman: SPE; 1992. SPE-24179-MS

[48] Schramm LL. Surfactants: Fundamentals and Applications in the Petroleum Industry. Cambridge University Press; 2000

[49] Gao Q, Yu X, Wang C, Ge Y, Zhang L, Li L, et al. Effect of oil displacement agent on corrosion behavior of NaCl-alkali/surfactant/polymer flooding pipeline steel. Vacuum. 2023;**209**:111742

[50] Kassa AM, Gasda SE, Kumar K, Radu FA. Impact of time-dependent wettability alteration on the dynamics of capillary pressure. Advances in Water Resources. 2020;**142**:103631

[51] Hammond PS, Unsal E. Spontaneous and forced imbibition of aqueous wettability altering surfactant solution into an initially oil-wet capillary. Langmuir. 2009;**25**(21):12591-12603

[52] Jadhunandan P, Morrow NR. Effect of wettability on waterflood recovery for crude-oil/brine/rock systems. SPE Reservoir Engineering. 1995;**10**(01):40-46

[53] Johannesen EB, Graue A. Mobilization of remaining oil-emphasis on capillary number and wettability. In: SPE International Oil Conference and Exhibition in Mexico. Veracruz, Mexico: SPE; 2007. SPE-108724-MS

[54] Zhao X, Blunt MJ, Yao J. Pore-scale modeling: Effects of wettability on waterflood oil recovery. Journal of Petroleum Science and Engineering. 2010;**71**(3-4):169-178

[55] Noruzi Y, Sharifi M, Fahimpour J, Sabet M, Akbari M, Hosseini S. The state-of-the-art of wettability alteration in sandstones and carbonates: A mechanistic review. Fuel. 2024;**356**:129570

Chapter 3

Preferential Solubilization of Fragrances in Micelles with Different Geometric Shapes

Vera Tchakalova

Abstract

Surfactant self-assembled aggregates, the main components of consumer products, offer a solution for the solubilization of fragrances, which is crucial to the consumer's product choice. The interactions between surfactant aggregates and fragrances are complex: surfactants influence fragrance release and performance, whereas fragrances affect the macroscopic properties of the formulation by changing the aggregate's shape and size. The present chapter describes studies on the solubilization of some fragrance ingredients in spherical and cylindrical micelles for a better understanding of their influence on micellar structure, viscosity, conductivity, and solubilization capacity. Turbidity, conductivity, and viscosity measurements were performed simultaneously in order to monitor the solubilization of fragrance molecules and the geometric transition of the surfactant aggregate.

Keywords: solubilization, turbidity, viscosity, maximum additive concentration, perfumery ingredients, micelles

1. Introduction

Most home and personal care consumer products, such as detergents, all-purpose cleaners, dishwashing products, shower gels, shampoos, and liquid soaps, represent aqueous surfactant solutions with different surfactant concentrations that are, in general, composed of a mixture of anionic and nonionic surfactants. All of these products solubilize fragrances—complex olfactive mixtures of volatile ingredients—at different concentrations, depending on the application field. Consumers are highly sensitive to the olfactive profile, and often, the fragrance is the predominant factor in the choice of a given product. In many countries (in South America, for example), the olfactive profile and fragrance intensity are the main criteria for the perception of cleanliness and personal hygiene.

However, most perfumery compounds are strongly hydrophobic, hence possessing low solubility in water. Their hydrophobic character hampers their incorporation in water-based fragrance formulations. Amphiphilic association structures such as surfactant molecular aggregates offer a solution to the solubilization problem of oily synthetic perfumes and have been the subject of intensive investigation over the last

IntechOpen

decade [1–6]. Scientific interest has been stimulated by subtle issues arising from the understanding of the solubilization phenomenon (e.g., the chemical nature and hydrophobicity of the perfume itself [1, 2, 4], the influence of the surfactant structure and its properties [1, 3, 7, 8], the perfume location in surfactant aggregates [5, 9–14], or the interfacial curvature change of the surfactant assembly with the addition of perfume [8, 15, 16]). The fragrance ingredients represent small molecules (MW < 500 g/mol) that often combine two or more chemical functionalities, making them powerful "troublemakers" or efficient "helpers" in aqueous surfactant solutions. They can induce change in product properties such as turbidity, phase separation, viscosity, and others. The fragrance ingredients are distributed in the surfactant palisade layer (at the hydrophobic/hydrophilic interface) or the hydrocarbon part of the surfactant's aggregates in different proportions, depending on the surface affinity and the level of hydrophobicity. In a previous study [17], we attempted to classify the fragrance ingredients in surface activity by using the parameter $EACN_{mix}$ (equivalent alkane carbon number for mixture) to express the relative fragrance interfacial activity with respect to a reference oil. We established a rule according to which fragrance ingredients with an $EACN_{mix}$ of <5.5 possess surface activity and preferential partitioning into the palisade surfactant layer of the droplets. Thus, fragrance ingredients having an $EACN_{mix}$ of >5.5 are highly apolar and prefer to be localized in the micellar core.

The shape, size, charge, and diffusivity of the surfactant micelles also have an important role in successful fragrance solubilization. The most frequent practical problems are undesired product turbidity and viscosity changes induced by fragrance solubilization even at relatively low fragrance concentrations (0.25 to 1%wt in personal and home care products and up to 2%wt in some fragrance boosters). Fragrance content is limited not only by legislation but also by solubilization. Therefore, studies of fragrance solubilization mechanisms, rates, and limits, along with their consequences on the macroscopic properties of a product, are valuable for the consumer goods industry.

The solubilization limit, characterized by the so-called Maximum Additive Concentration (MAC), is a measure of the solubilizing capacity of surfactant aggregates for a given solute. The value of the MAC is affected by and mainly dependent on parameters such as the surfactant assembly itself and the hydrophobic character of the guest molecule, as well as temperature, pH, and electrolytes. Numerous studies have reported the maximum solubilization of fragrance ingredients in solutions containing different types of surfactants (nonionic, anionic, and cationic) and mixtures of [1–4], focusing on the type of surfactant and not on the surfactant aggregate shape. Nonetheless, few studies have answered the following questions:

1. What is the solubilization capacity of surfactant aggregates with different geometries for fragrance molecules with different functionalities?

2. Which micelle geometry must be chosen for the specific solubilization of a fragrance with known hydrophobicity/surface activity?

The answers to these questions could help in the creation of compatible (surfactant self-assembly architecture–fragrance) combinations while avoiding technical issues.

The solubilization capacity of rod-like surfactant aggregates for different additives was investigated by Hoffmann and Ulbricht [18]. They studied the solubilization of hydrocarbons (linear and aromatic), linear alcohols, and esters (with different chain lengths) in aqueous surfactant solutions of tetradecyltrimethylammonium bromide (cationic surfactant) in the presence of equimolar quantities of sodium salicylate. They reported that linear alkanes at a low hydrocarbon-to-surfactant ratio induced a drastic decrease in light scattering, indicating a transformation from rod-like to spherical micelles. The aromatic hydrocarbons did not always show the same behavior at low concentrations (<10 mM). The authors mentioned that the effect was correlated to the chain length of the molecules. The alkanes could induce the formation of a coacervate phase, and the transition from rod to sphere could be observed at a higher solute concentration (>20 mM). The authors also studied the behavior of alcohols and demonstrated that those with a chain length less than that of n-pentanol reduced the rods to spheres, whereas the longer chain length alcohols stabilized the rods. The transformation of rods to spheres has been demonstrated at low concentrations (up to 15–20 mM) for both ionic (cationic) and nonionic micelles.

The esters behaved more as hydrocarbons than as alcohols. For example, ethylhexanoate had a similar effect as n-hexane. The esters can be solubilized in a relatively high concentration in the rod-like micelles before reducing the size of the rods. Thus, in contrast to that of alkanes, the solubilization of alcohols in surfactant self-assemblies is enhanced when the micelles are spherical, whereas cylindrical micelles allow the solubilization of alkanes better than they do of alcohols. The authors explain the observed phenomenon as an entropic gain due to geometric constraints: the required interfacial area per molecule in spherical micelles is higher than in rod-like micelles, but the hydrophobic core volume is smaller. Therefore, the spherical micelle allows better solubilization of surface-active molecules such as some alcohols and poorer solubilization of highly hydrophobic molecules such as hydrocarbons. Rod-like micelles can accommodate more hydrophobic molecules in the core, but the interfacial area is closely packed, which avoids almost any contact between the water and the hydrocarbon and lacks space for the cosurfactants.

At higher concentrations, the behavior of the short alcohols, solubilized in cylindrical micelles, was shown to be different: their solubilization caused the appearance of lamellar phases or the coexistence of a micellar solution and a lamellar phase. In this case, the authors distinguished the I-type of lamellar phase, which is formed mainly as a result of interactions between the surfactant and the cosurfactant and possesses a high solubilization capacity.

All of these micellar changes induced by the solubilization of solutes with different hydrophobicities and surface activities are important for industrial applications because they cause considerable viscosity and turbidity changes. The reduction in the size of the rod-like micelles would tend to decrease the solution viscosity, whereas the increase in the size of the micelles and the micelle-to-liquid-crystalline phase transition would increase the solution viscosity. For this reason, it is important to verify whether the same effects are observed with more complex molecules such as fragrance ingredients.

Perfumery ingredients cannot simply be classified in groups of hydrocarbons, alcohols, or esters. Often, they have a mixed structure composed, for example, of terpene and alcohol, or a combination of phenol and alcohol or aldehyde. In addition, in general, the concentrations used in consumer goods are higher than those applied in the previously mentioned studies (varying from 0.1 to 1%wt and exceptionally to 2% wt, which corresponds to a range of ∼4 to ∼150 mM). Therefore, it is not obvious that fragrance solubilization in spherical and cylindrical micelles would follow the rules established for linear hydrocarbons, alcohols, and esters.

Here, we report our studies on the solubilization of fragrance ingredients in spherical and cylindrical micelles, in which we aimed to determine the solubilization rate and solubilization capacity of micelles for these complex molecules. The effects of solubilization on the macroscopic properties, viscosity and conductivity, were also investigated. The correlation between the interfacial activity of perfumery ingredients and turbidity and viscosity is discussed. We chose to use a simple solubilizing system, composed of the anionic ethoxylated surfactant sodium lauryl ether sulfate (SLES, EO2), which is the main surfactant in rinse-off products [19]. The data were obtained with an experimental setup designed specifically for the investigation of solubilization capacity or the elaboration of phase diagrams [20]. Turbidity, conductivity, and viscosity measurements were performed to determine the solubility of the fragrances and the phase transition borders in case of surfactant aggregate shape evolution. Polarized microscopy was applied to confirm the existence of liquid crystalline phases.

We believe that this study would be useful in providing guiding principles for the formulation of successful fragranced surfactant-based products.

2. Experimental section

2.1 Materials

The basic formulation in many personal care products consists of a mixture of anionic and nonionic surfactants in water [21]. For the anionic surfactant, we used SLES with two ethoxylated groups on average (Texapon® N70, Cognis GmbH, Germany).

The perfumery ingredients phenethyl alcohol, eugenol, linalol, lilial®, hexyl salycilate, and limonene are high-purity products (>98%) supplied by Firmenich SA. Their chemical structures and hydrophilicity/hydrophobicity classification are shown in **Table 1**. As can be seen from the log P or $EACN_{mix}$ values, the hydrophilic properties of the perfumery ingredients follow the order phenethyl alcohol > eugenol > linalol > lilial > limonene > hexyl salycilate.

Fragrance ingredient		Functionality	log P[a]	$EACN_{mix}$ [a]
Phenethyl alcohol	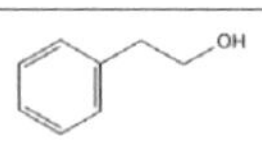	Aromatic alcohol	1.41	1.96
Eugenol	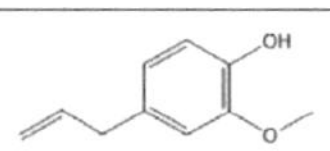	Phenylpropene	2.14	2.99
Linalol		Terpene alcohol	2.94	4.33
Lilial	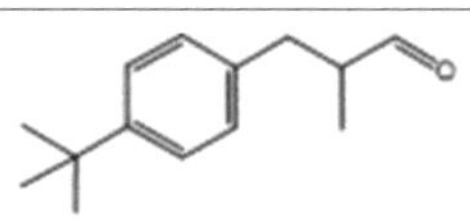	Aromatic aldehyde	3.90	5.70
Limonene	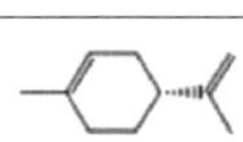	Cyclic terpene	4.38	6.00
Hexyl salicylate	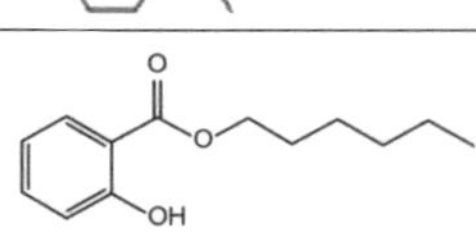	Phenyl ester	5.06	6.39

[a]*The values of log P and $EACN_{mix}$ are given in Ref. [19].*

Table 1.
Chemical functionality, log P, and $EACN_{mix}$ of the different perfumery raw materials used in this study.

2.2 Experimental procedure

In this study, a multiparameter scan analysis method was used to obtain the maximum solubilization capacity of micellar surfactant phases for different fragrance molecules. This analysis method allows the simultaneous determination of different physicochemical parameters such as transient turbidity, viscosity, or conductivity after a concentration jump during titration. The experimental setup, "Multi Parameter Scanning Instrument v.2 (MPS-2)," supplied by Scanalys® (Sweden), is schematically presented in **Figure 1**. The working principle is to measure and record several parameters at a time, while controlling the temperature and composition. This setup combines several devices (controllers) to measure transient conductivity, pH, turbidity, and viscosity. The inspection window with a polarizer filter allows visualization of the formation of the liquid crystalline phase during the experiments. The precise description of the experimental setup is given in Ref [20].

Small quantities of a perfumery ingredient were added to the continuously stirred surfactant system. The titration was performed by steps of 0.25%wt fragrance ingredient concentrations with automatic syringes. The time between each fragrance ingredient addition was set to 10 min. This time was sufficient for the complete solubilization of each portion of 0.25%wt up to high concentrations. The experiments were performed at a temperature of 25°C. Turbidity and conductivity were measured continuously every 5 s. These two parameters measured at the same time during solubilization allowed us to know immediately and precisely the effect of fragrance on the surfactant system.

2.2.1 Conductivity

Conductivity, measured simultaneously with turbidity, was used to determine the interfacial curvature change of the surfactant structure or the presence of a two-phase solution. On the one hand, a phase transition of the surfactant aggregates from micellar to a more organized structure (such as hexagonal or lamellar phase) would reduce the mobility of the counterions of the charged surfactants and hence would tend to decrease the conductivity of the solubilizing surfactant structure. On the other hand, above the solubility limit, the water-based solution will be in equilibrium with an

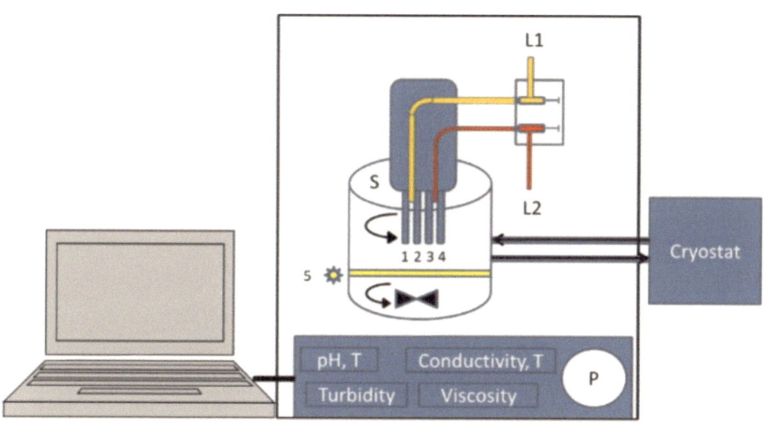

Figure 1.
Schematic presentation of scanning instrument MPS-2. L1 and L2 indicate the automatic syringes; 1–4 correspond to the different measurement devices; 5 is the light beam for turbidity measurement; S means sample, and P represents the window with polarization filters.

excess oil phase whose conductivity is lower than that of the water. As a result, a decrease in the overall solution conductivity in combination with a strong increase in turbidity would allow determination (confirmation) of the MAC.

2.2.2 Rheology

The viscosity measurement was used to detect the change in shape and size of the micelles, as well as the phase transitions, which can occur during the solubilization of a perfumery ingredient into surfactant aggregates. The viscosity of each fragrance/surfactant system was measured on a C-VOR 150 rheometer from Bohlin Instruments by using the cone-plate technique. The measurements were performed with an increasing shear rate from 0.01 to 100 s^{-1}. The plate temperature was controlled with a Peltier unit at 25°C.

The viscosity of the water/surfactant/perfumery ingredient is mainly governed by the viscosity of the continuous phase, that is, the surfactant phase in the aqueous solution [16, 22]. In addition, the rheological behavior of the different surfactant aggregates is variable [23].

3. Results and discussion

3.1 Formulation and identification of surfactant aggregates

From our previous studies and from the literature [24], we know that at low concentrations, SLES forms spherical (or globular) micelles. With increasing surfactant concentration, they transform into cylindrical micelles, followed by hexagonal phases and finally lamellar phases. In order to determine the phase boundaries precisely, we plotted the viscosity as a function of surfactant concentration at a fixed shear rate (52 s^{-1}) and temperature (25°C) (**Figure 2**). The slope change on the viscosity curve corresponds to the phase boundaries. When SLES forms spherical

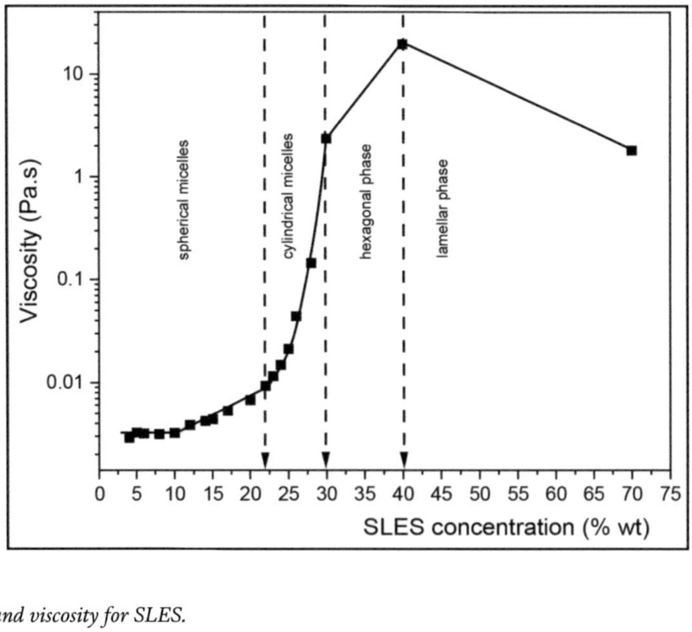

Figure 2.
Phase behavior and viscosity for SLES.

micelles, viscosity remains unchanged. Above 23%wt SLES, viscosity increases due to a phase transformation in rod-like micelles. Above 28%wt SLES, the cylindrical micelles are forced to pack into a hexagonal phase and the viscosity is superior to 10 Pa.s. Finally, above 45%wt SLES, lamellar phases are formed with a viscosity lower than that of the hexagonal phase.

For the purpose of our study, we chose to work with the concentrations 10%wt SLES (spherical micelles) and 23%wt SLES (cylindrical micelles) for the single surfactant system [25].

3.2 Solubilization of fragrance ingredients in spherical and cylindrical micelles: general trends

We investigated the solubilization of six fragrance ingredients (**Table 1**) with different hydrophobicities and surface activities into SLES aggregates with two different micellar geometries: spherical (or globular) and cylindrical micelles. Turbidity, viscosity, and conductivity as a function of the fragrance ingredient concentration are presented in **Figures 3** and **4** for all ingredients solubilized in 10%wt

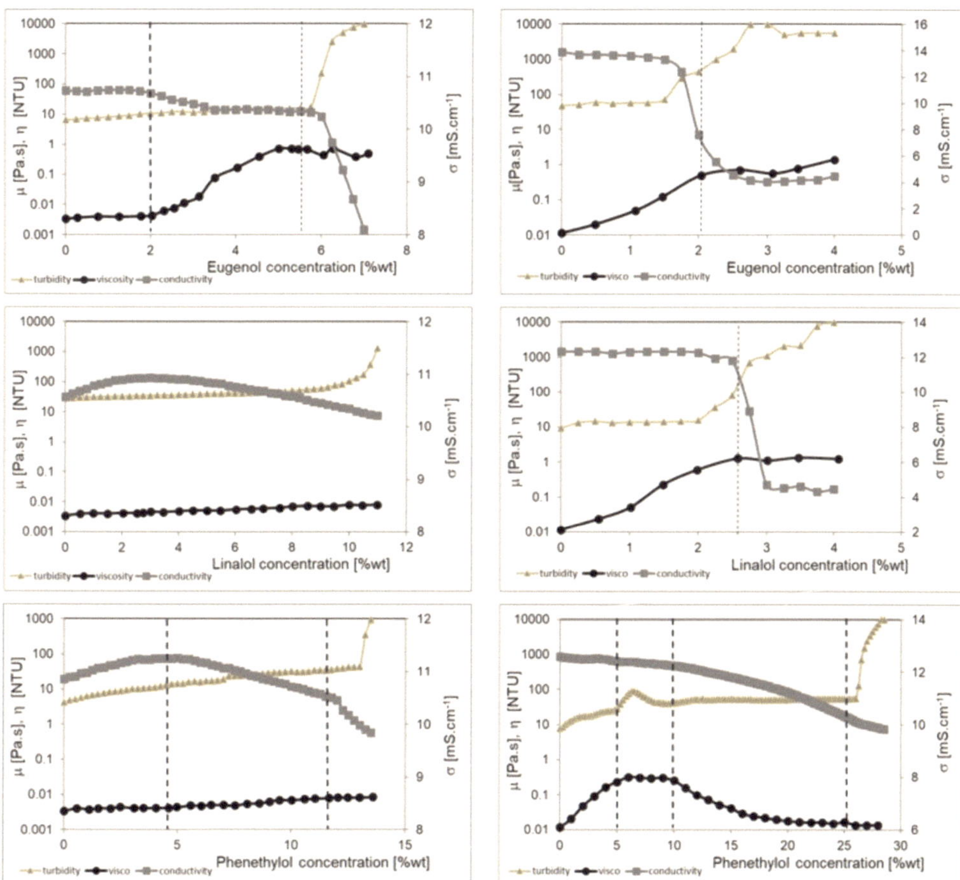

Figure 3.
Turbidity (gold symbols), viscosity (black symbols), and conductivity (gray symbols) of surfactant solutions containing 0.25%wt perfumery ingredient, (A) eugenol, (B) linalol, (C) phenethylol, and 10%wt SLES (left column) and 23%wt SLES (right column).

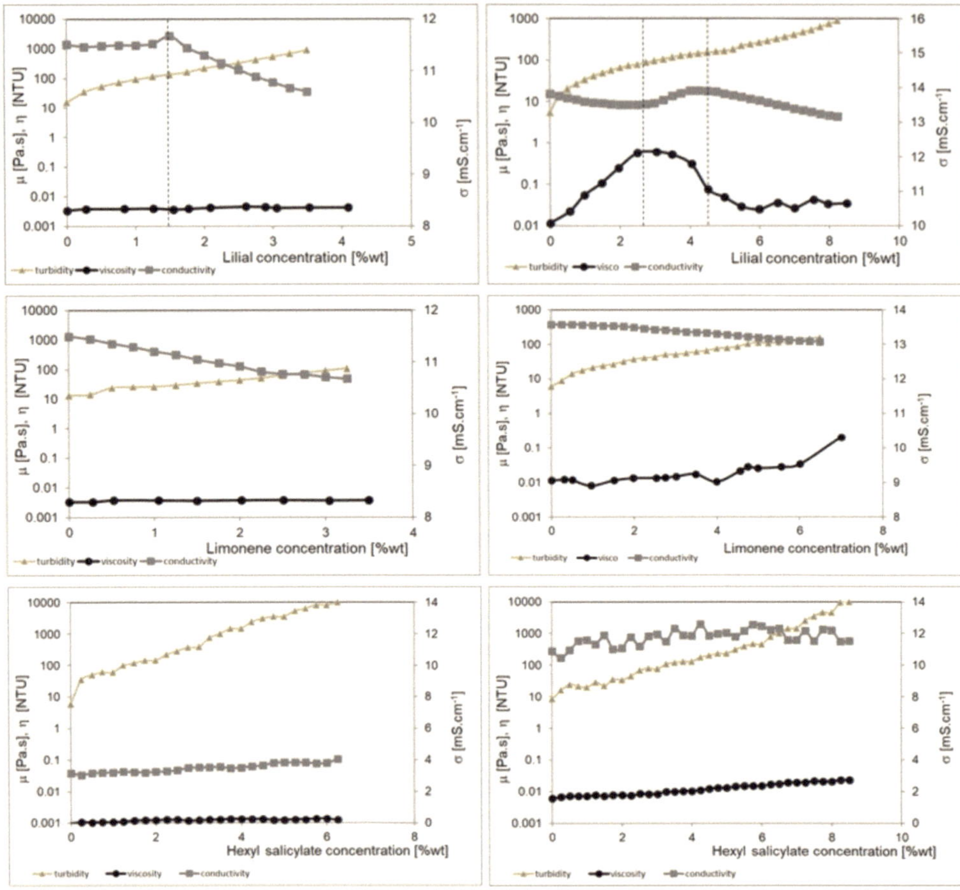

Figure 4.
Turbidity (gold symbols), viscosity (black symbols), and conductivity (gray symbols) of surfactant solutions containing 0.25%wt perfumery ingredient, (A) lilial, (B) limonene, (C) hexyl salicylate, and 10%wt SLES (left column) and 23%wt SLES (right column).

Fragrance ingredient	10%wt SLES		23%wt SLES	
	Cr. Conc (%)	Transition	Cr. Conc (%)	Transition
Eugenol	2.0	Sph - Cyl	2.0	Cyl - Hex
	5.5	Cyl - Lam		
Linalol	3.0		2.5	Cyl - Hex
Phenylethyl alcohol	—	—	5.0	Cyl - Lam
	—	—	10.0	Lam - Cyl
Lilial	—	—	4.0	Cyl - Lam

Table 2.
Structural transitions of micellar phases induced by fragrance ingredient solubilization.

SLES (left column) and 23%wt SLES (right column). For each ingredient, the changes in turbidity, viscosity, and conductivity could be compared as a function of the fragrance concentration.

The morphological transitions induced by the solubilization of the fragrance ingredients in the different micellar solutions are summarized in **Table 2**. As shown, the fragrance alcohols strongly influenced the micellar aggregates by changing their geometry and, consequently, the viscosity of the solutions.

Eugenol induced two different aggregation changes at 2 and 5.5%wt in the 10%wt SLES solution. These changes could be observed as slope changes on the conductivity and viscosity curves, whereas turbidity increased continuously. The first transition should correspond to the spherical to cylindrical micelle transition and the second to the cylindrical to lamellar phase transition. The lamellar phase was identified from the texture on the microscopic images taken under polarized light. Solubilized in cylindrical micelles (23%wt SLES solutions), eugenol led to the formation of a hexagonal phase at 2%wt concentration. This fragrance ingredient is highly surface active with a very strong influence on the interfacial curvature of the microemulsion droplets in both spherical and cylindrical micelles.

Linalol solubilization led to a continuous increase in turbidity and viscosity in the solutions with spherical micelles. Conductivity at low linalol concentrations slightly increased and then started to decrease progressively. Phase transitions were not observed. The influence of linalol on the interfacial curvature seems less important than that with eugenol. In the solutions with cylindrical micelles, the effect of linalol solubilization was stronger. A clear phase transition from cylindrical micelles to the hexagonal phase was observed. The hexagonal phase was identified by microscopy in polarized light. Turbidity increased significantly, and viscosity maintained a constant high value, a characteristic property of the hexagonal phase.

The solubilization of phenylethyl alcohol in the solutions with spherical micelles induced a progressive, slight increase in viscosity and turbidity that was related to the change in micelle size. The conductivity curve went through a maximum value and then started to decrease similar to the curve for linalol. Linalol and phenylethyl alcohol could be solubilized up to very high concentrations in the spherical micelles without a significant perturbation of the macroscopic properties of the solutions. However, solubilization of phenylethyl alcohol in the cylindrical micelles (at 23% wt SLES) induced a couple of transitions at 5 and 10%wt fragrance concentrations. The turbidity and viscosity curves both underwent a maximum. At lower concentrations (<5%wt phenylethyl alcohol), a transparent viscous solution was obtained, which should contain cylindrical micelles. In the range from 5 to 10%wt phenylethyl alcohol, an intermediate lamellar phase was formed, and at concentrations >10%wt phenylethyl alcohol, an isotropic transparent gel was obtained until the critical concentration of 26%wt phenylethyl alcohol, at which point a phase separation was achieved.

The solubilization of lilial in the spherical micelles, observed on the turbidity and the viscosity curves, did not induce strong changes in the surfactant aggregate shape. However, the conductivity curve changed the shape strongly: from a constant value at the low concentrations of lilial, it started to decrease progressively above 1.5%wt lilial, indicating some charge or ion mobility reduction. The absence of a liquid-crystalline phase and the high solution turbidity, combined with the strongly decreasing conductivity, indicated a phase separation in microemulsion in equilibrium with an oil excess phase. Integrated into the cylindrical micelles (at 23%wt SLES), lilial induced different transitions. The conductivity curve (**Figure 4A**, right) is more complex than the curve corresponding to the spherical micelles, containing a minimum at low lilial concentrations and a maximum at higher lilial concentrations. The minimum at 3%wt lilial, indicating the branching of the cylindrical micelles, correlated very well with the viscosity peak. At above 5%wt lilial in the cylindrical micellar phase, the system turned into a lamellar phase. The value of the conductivity of the lamellar phase was higher compared with that of the hexagonal phase, obtained at 2.5%wt eugenol in cylindrical micelles. This difference between both types of liquid crystalline phases is well correlated to their viscosity and the mobility of the free electrolyte ions in the solution. The turbidity increased continuously without abrupt changes.

The solubilization of limonene and hexyl salicylate increased the turbidity continuously and the viscosity slightly without changing the morphology of the phase and the micelle shape at both surfactant concentrations. The conductivity kept an almost constant value in the whole range of fragrance ingredient concentrations. Limonene is an aromatic hydrocarbon and did not transform the cylindrical micelles into spherical ones, but rather stabilized the cylinders. This effect is in agreement with the studies of Hoffmann and Ulbricht [18] concerning aromatic alkanes. Hexyl salicylate is a phenyl ester and, as was found by Hoffmann and Ulbricht for the esters, it behaved similar to limonene without induction of any micelle shape transition or morphological change. We did not observe the reduction in turbidity (light scattering in [18]) reported by Hoffmann and Ulbricht due to a transformation of cylindrical to spherical micelles observed during the solubilization of the alkanes, short-chain alcohols, and esters in the range of their concentrations (up to 20 mM) and even at higher concentrations. Possible reasons for the disparity in results could be a) the mixed functionalities of the fragrance molecules or b) the starting high concentration (0.25% > 10 mM). To clarify the reasons for the different observations, we performed experiments in which hexane and decane were solubilized in the cylindrical micelles (23%wt SLES), expecting to reproduce the transformation effect. However, only a continuous increase in the turbidity and viscosity was observed (**Figure 5**).

Comparison of the turbidity of the solutions containing hexane, decane, and hexyl salicylate in cylindrical micelles (23%wt SLES) demonstrated that hexyl salicylate solubilization is similar to that corresponding to decane solubilization. The inset in **Figure 5** shows the behavior at low concentrations (<1%wt oil). Some reduction of the turbidity could be observed for the three oils, but the effect was negligibly weak

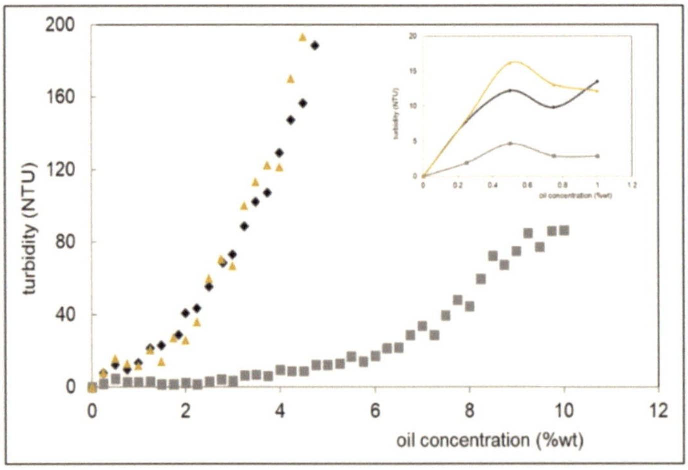

Figure 5.
Turbidity as a function of the oil concentration of solutions at 23%wt SLES, solubilizing hexane (gray symbols), hexyl salicylate (gold symbols), and decane (black symbols). Inset: Turbidity as a function of oil concentration (<1%wt oil).

and did not induce viscosity changes. Therefore, the transformation should happen only at very low concentrations or it should take longer than our experimental time, as was reported in [26] for the solubilization of triolein in cylindrical micelles of nonionic surfactant (3 to 16 days).

3.3 Solubilization rate and MAC

3.3.1 Turbidity (η)

The solubilization capacity, expressed by the MAC, was defined as the highest concentration of fragrance molecules that could be incorporated into the micellar structure at a given surfactant concentration before the two-phase separation. A typical experimental turbidity curve as a function of time and fragrance ingredient concentration is presented in **Figure 6**. The sequence of turbidity peaks corresponds to the stepwise addition of the fragrance ingredient [1, 3, 4]. When the solubilization limit is reached, a strong asymptotic and irreversible increase in turbidity takes place [20, 27], as shown in **Figure 6**, revealing the coexistence of two phases (i.e., excess oil phase in equilibrium with a microemulsion or liquid crystalline phase). The separation of the perfume-surfactant system in two phases is confirmed by centrifugation. This critical fragrance ingredient concentration is considered the MAC of the current surfactant system.

The turbidity of the surfactant solution before the fragrance addition is denoted as η_0. As soon as the perfumery ingredient is injected, the turbidity jumps strongly because of the spontaneous formation of a two-phase solution. During the process of mixing and solubilization, the turbidity exponentially decreases and reaches a constant value B, as depicted in the inset for **Figure 6** [27]. After each titration of a perfumery ingredient, the turbidity plateau B could theoretically recover the initial value η_0 of the surfactant solution (obtained before the turbidity peak), indicating the absence of any structural or size change in the solubilizing surfactant aggregate.

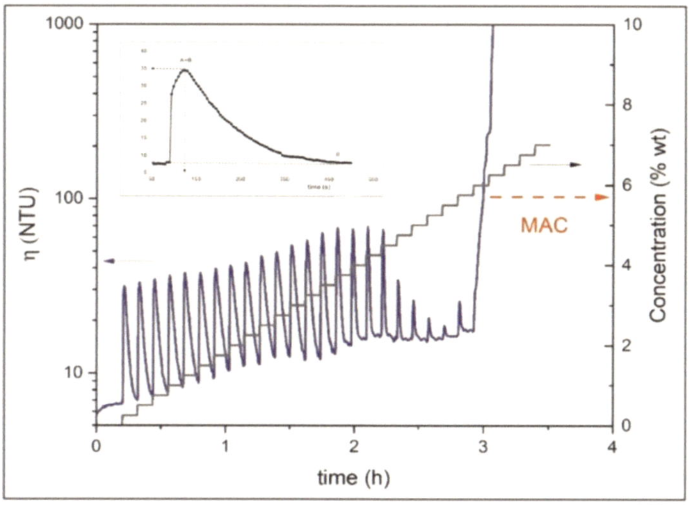

Figure 6.
Turbidity (η) experiments on a 10%wt SLES solution (η^o) in which successive amounts of a fragrance ingredient (0.25%wt) are added. (inset) turbidity change during the solubilization of 0.25%wt fragrance.

In practice, oil solubilization always leads to swelling of the micelles and a consequent increase in their size, and as can be observed from **Figure 6**, the turbidity plateau B reaches a value higher than η_0, indicating an increase in micelle size or number.

The turbidity decay as a function of time (**Figure 6** inset) was fitted by using the following model:

$$\eta(t) = Aexp\left[-\tau\left(t - t_{peak}\right)\right] + B \tag{1}$$

where η is the turbidity (NTU), τ is the solubilization rate (s^{-1}), and A and B are constants. At $t \rightarrow \infty$, $\eta \rightarrow B = \eta_s$, and at $t = t_{peak}$, $\eta_{max} = A + B$.

If the turbidity results are presented in the form:

$$\left[\frac{\eta - \eta_s}{\eta_{max} - \eta_s}\right] = exp\left[-\tau\left(t - t_{peak}\right)\right], \tag{2}$$

the solubilization rate τ (s^{-1}) can be determined as the curve slope, assuming that η, η_s, η_{max}, and t_{peak} are experimental values for all molecules.

The constant η_s is the turbidity value at the end of the solubilization process of each injected fragrance quantity. If the solubilization does not induce any change in the micelles, the value of η_s should be equal to that before the injection of the fragrance (η_0). Increasing the solute concentration causes the micelles to swell and increase in size, which should lead to an increase in turbidity. Therefore, the plateau value η_s should be directly related to the droplet radius and density (interactions).

As can be observed from **Figure 7**, the solubilization rate depends not only on the type of micelles but also on the type of perfumery ingredients. The solubilization rate values (**Table 3**) required to solubilize 0.25%wt fragrance ingredient seem to be significantly sensitive to the perfume type in the spherical surfactant aggregate geometry. In contrast, in cylindrical micelles, the solubilization rate has similar and low values for all perfumery ingredients.

According to the theory of solubilization [28], the solubilization rate should be dependent on the oil properties *via* the parameters V_{oil} (oil molecular volume), D_{oil} (diffusion coefficient in the surfactant solution), and C_{eq} (oil solubility in water): $\beta = V_{oil} \times D_{oil} \times C_{oil}$ (**Table 4**).

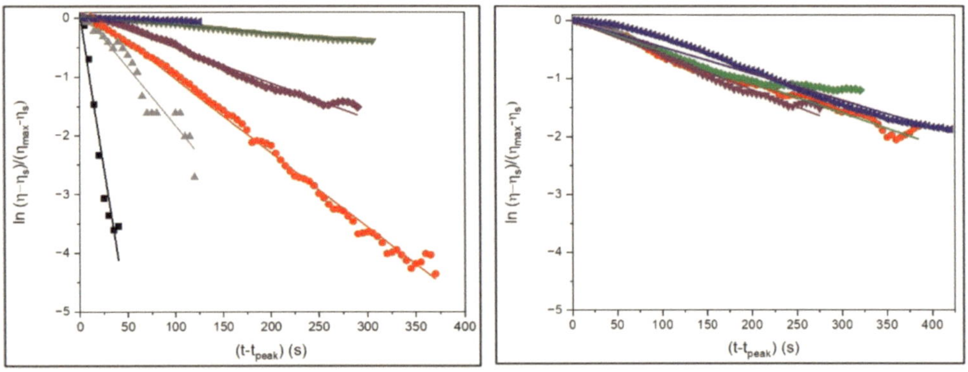

Figure 7.
Normalized turbidity as a function of time to solubilize 0.25%wt perfumery ingredient in (A) spherical (10%wt SLES) and (B) cylindrical (23%wt SLES) micelles: Phenethyl alcohol (black), eugenol (red), linalol (gray), lilial (green), limonene (violet), hexyl salicylate (blue).

Fragrance ingredient	τ (s^{-1}) (in spherical micelles)	τ (s^{-1}) (in cylindrical micelles)
Phenethyl alcohol	0.1020	0.003
Eugenol	0.0120	0.005
Linalol	0.0170	0.004
Lilial	0.0010	0.004
Limonene	0.0060	0.006
Hexyl salicylate	0.0004	0.004

Table 3.
Solubilization rate of the different perfumery raw materials studied in spherical and cylindrical micellar phases of SLES (10%wt SLES and 23%wt SLES, respectively).

Fragrance ingredient	Mvol (cm^3/mol)	D (cm^2/s)	Wsolubility (mg/L)	β (cm^2/s)
Phenylethyl alcohol	119.8	1.35E-05	21990.00	2.92E-07
Eugenol	154.9	1.05E-05	754.00	7.45E-09
Limonene	179.3	9.05E-06	683.70	7.20E-09
Linalol	217.3	7.47E-06	7.86	6.24E-11
Hexyl salicylate	162.1	1.00E-05	4.58	5.46E-11
Lilial	215.7	7.52E-06	6.01	4.39E-11
Decane	194.9	8.33E-06	0.05	5.93E-13

Table 4.
Maximum additive concentration of the different perfumery raw materials studied in spherical and cylindrical micellar phases of SLES (10%wt SLES and 23%wt SLES, respectively).

At the end of the solubilization process of each fragrance portion, the value of η_s was determined and presented as a function of the fragrance concentration. From this curve, the value of the MAC was found for each perfumery ingredient as the concentration at which the turbidity slope increases asymptotically. Therefore, this is the maximum fragrance concentration that can be incorporated in the current micellar system. The values of MAC are reported in **Table 5**. The results indicate that solubilization into spherical micelles is highest for phenethyl alcohol, followed by linalol and

Fragrance ingredient	MAC [%wt]	
	10%wt SLES	23%wt SLES
Phenethyl alcohol	13.1	4.7
Eugenol	6.0	2.3
Linalol	10.7	3.2
Lilial	2.4	6.4
Limonene	2.2	3.7
Hexyl salicylate	3.1	6.7

Table 5.
Maximum additive concentration of the different perfumery raw materials studied in spherical and cylindrical micellar phases of SLES (10%wt SLES and 23%wt SLES, respectively).

eugenol and then hexyl salicylate, lilial, and limonene. In the cylindrical micellar phase, the order of the MAC values changes: hexyl salicylate is the most solubilized, followed by lilial, phenethyl alcohol, and limonene and then linalol and eugenol. These results clearly indicate preferential solubilization of the fragranced molecules with surface activity in surfactant aggregates with spherical geometry. The highly apolar fragrance molecules are preferentially solubilized by the cylindrical surfactant aggregates. Our results confirmed the preferential solubilization effects observed in [18] for the perfumery ingredients.

Notably, increasing the surfactant concentration in the aqueous solution is not a sufficient factor to increase the MAC value, in particular for fragrance alcohols (such as eugenol and linalol).

In our previous studies [17], we introduced the parameter $EACN_{mix}$, a key indicator for the interfacial solubility of the perfumery ingredients and mixtures. We found that fragrance molecules leading to a low $EACN_{mix}$ value, $EACN_{mix} < 5.5$, have an affinity for the interfacial domain within the surfactant aggregates, whereas those having a higher $EACN_{mix}$ value, $EACN_{mix} > 5.5$, are solubilized into the apolar core of the surfactant structures. The correlation of the solubilization rate τ and MAC to the $EACN_{mix}$ parameter for the perfumery molecules is presented in **Figure 8A** and **B**. The curves indicate that perfumes with high surface activity such as phenethyl alcohol, eugenol, and linalol ($EACN_{mix}$ = 1.96, 2.99, and 4.33 respectively; see **Table 1**) are solubilized faster and at higher quantity than the highly hydrophobic perfumes in the spherical micelles. The cylindrical micelles are preferred by the three hydrophobic perfumery ingredients, lilial, limonene, and hexyl salicylate ($EACN_{mix}$ = 5.7, 6.0, and 6.39, respectively; see **Table 1**). The observed phenomenon is closely related to the results of the previous studies concerning the correlation between the packing parameter, the interfacial concentration of the fragranced molecules, and the $EACN_{mix}$ values [17, 29].

For fragrance alcohols, it was commonly observed [8, 13, 15, 30] that these molecules (including eugenol and linalol) act as cosurfactants and penetrate into the palisade layer of the surfactant self-assembly. As a consequence, the larger the hydrophilic region of the micelle, the higher the solubilization/incorporation of the fragrance molecule with surface activity between the surfactant head groups. For geometric reasons, as was clearly explained by Hoffmann and Ulbricht [18], the

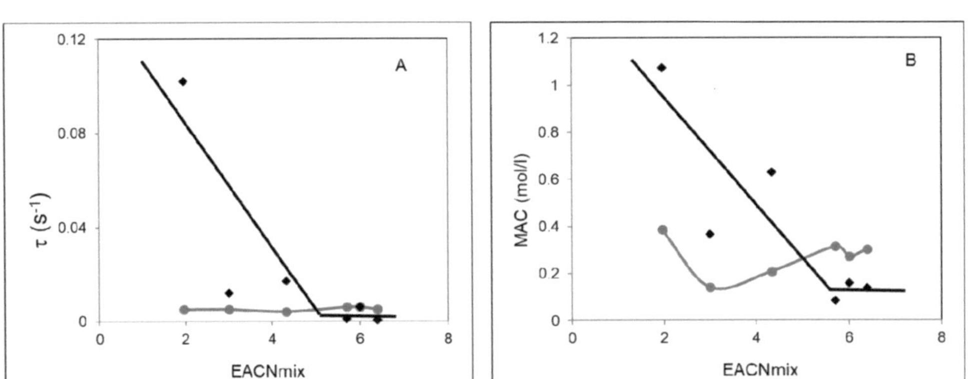

Figure 8.
Solubilization rate τ (A) and MAC (B) of perfumery ingredients in spherical (10%wt SLES) (♦) and cylindrical (23%wt SLES) (•) micelles as a function of $EACN_{mix}$.

spherical shape requires that the area per surfactant head group is $a_s \geq 3a_0$, where a_0 is the cross section of the hydrophobic surfactant chain. The rod-like shape requires that $a_r \geq 2a_0$. Therefore, the surface, accessible to the surface-active molecules, is higher in the spherical micelles. Moreover, the adsorption of fragrance molecules at the interface leads to a decrease in electrostatic repulsion between the SLES head groups and thus to closer packing of the palisade layer. This effect could also be observed as a decrease in conductivity. The consequence is transition from spherical to rod-like micelles [31]. For these reasons, phenethyl alcohol, eugenol, and linalol can be solubilized in higher quantities in spherical than in cylindrical micelles.

On the other hand, the highly hydrophobic perfumery ingredients such as hexyl salicylate and limonene, localized mostly in the micellar core, have a negligible influence on the packing parameter, mostly by oil penetration, between the hydrophobic surfactant chains. As is mentioned in [18], the solubilization of non-surface-active hydrophobic molecules in spherical micelles is limited to the micellar core and leads to swelling of the micelle without important curvature changes. The micelles swell until the limit of the packing surface area is reached, and the addition of more fragrance molecules induces a two-phase separation.

The relative turbidity, equal to the ratio Bi-η_0/η_0 after the solubilization of each portion i of fragrance and before reaching MAC, is presented in **Figure 9** as a function of the volume fraction of fragrance in the surfactant solutions of spherical (A) and cylindrical (B) micelles. The curves can be sufficiently well fitted by a linear model:

$$\frac{\eta_s - \eta_0}{\eta_0} = \alpha\Phi_o \tag{3}$$

where Φ_o is a perfumery ingredient volume fraction and α is a constant, determined by the curve's slope. The turbidity values considered here correspond to solubilization in a single phase (spheres or cylinders). Any phase transitions are excluded. Approaching the phase boundaries, the turbidity changed exponentially as a function of the perfumery ingredient volume fraction.

Turbidity is a measure of the light scattered by the aggregates dispersed in the aqueous solutions. The relationship between turbidity and the properties of the scattering system is defined by the following equation [32]:

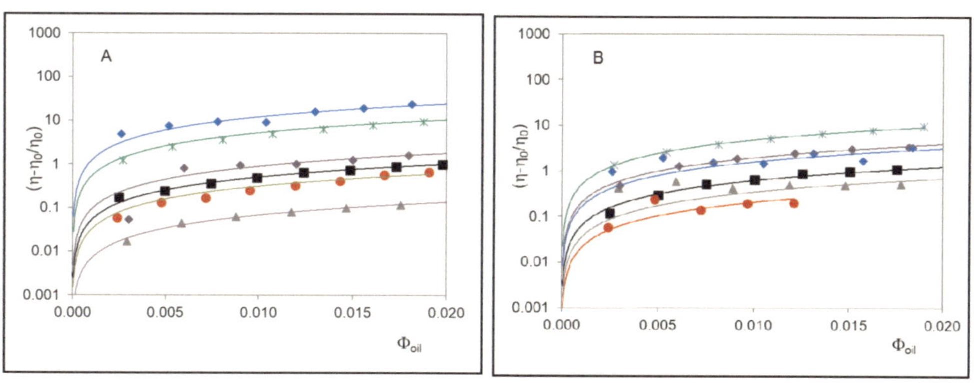

Figure 9.
Relative turbidity as a function of the fragrance volume fractions in SLES aqueous solutions: (A) 10%wt SLES and (B) 23%wt SLES. Symbols are experimental points, and lines are fitting curves with linear dependence: Phenylethyl alcohol (black), eugenol (red), linalool (gray), lilial (green), limonene (violet), hexyl salicylate (blue).

$$\eta = R_{\pi/2}\left[\frac{16}{3}\pi - \frac{128}{15}\frac{R^2\pi^3}{\lambda^2}\right] \tag{4}$$

where R is the droplet radius, λ is the light wavelength, and $R_{\pi/2}$ is:

$$R_{\pi/2} = \frac{2\pi^2 n_w^2}{\lambda^4}\left(\frac{dn}{d\phi}\right)^2 V_{sphere}\phi S(0) \tag{5}$$

where *n* is the refractive index, Φ is the volume fraction of the scattering objects (micelle + oil), S(0) is the structure factor, and V_{sphere} is the volume of droplets. These equations show that the turbidity is a function of micelle characteristics, as well as of oil characteristics directly *via* the refractive index increment $\frac{dn}{d\phi}$ and indirectly *via* the droplet size.

$$\frac{dn}{d\phi} = \frac{3}{2}\varepsilon_w^{1/2}\left(A\frac{\phi_o}{\phi} + B\frac{\phi_s}{\phi}\right) \tag{6}$$

ϕ_o and ϕ_s are the oil and surfactant volume fractions, respectively, and ϕ is the total volume fraction of the scattering particles, $A = (\varepsilon_o - \varepsilon_w)/(\varepsilon_o + 2\varepsilon_w)$ and $B = (\varepsilon_s - \varepsilon_w)/(\varepsilon_s + 2\varepsilon_w)$, with ε_i being the dielectric constants of oil, water, and surfactant, assuming that $n^2 = \varepsilon$.

Thus, α should be dependent on the oil properties directly *via* the refractive index increment (dielectric constant of the oil) and indirectly *via* the droplet's radius, both parameters closely linked to the hydrophobicity (polarity) and surface-active characteristics of the fragrance ingredients.

3.3.2 Viscosity (μ)

Figure 10 depicts the dependence of the specific viscosity on the oil volume fraction for all fragrance ingredients solubilized in (A) spherical micelles and (B) cylindrical micelles. The viscosity changes induced by fragrance solubilization depend on both the micellar system used and the type of fragrance ingredient, as can be observed from the graphs.

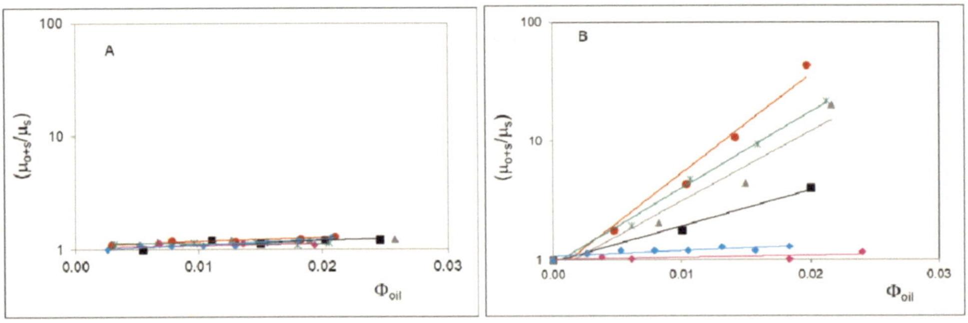

Figure 10.
Relative viscosity as a function of the fragrance volume fractions in SLES aqueous solutions: (A) 10%wt SLES and (B) 23%wt SLES. Phenylethyl alcohol (black), eugenol (red), linalool (gray), lilial (green), limonene (violet), hexyl salicylate (blue).

The correlation between the specific viscosity and the volume fraction of oil for the solutions at 10%wt SLES and at fragrance ingredient concentrations up to 2%wt (**Figure 10C**) follows a linear relation:

$$\frac{\mu_{o+s}}{\mu_s} - 1 = \gamma\Phi_0 \tag{7}$$

with coefficient γ being similar for all fragrance ingredients and having values in the range from 7 to 15. According to the slope coefficient value, the fragrance ingredients could be arranged in the following order: eugenol > linalol > phenylethyl alcohol> hexyl salicylate > lilial > limonene.

For the solubilization in the cylindrical micelles (at 23%wt SLES), the correlation between the specific viscosity and the volume fraction of oil (**Figure 10D**) cannot be fitted with linear relation but follows an exponential relationship:

$$\frac{\mu_{o+s}}{\mu_s} - 1 = \exp(\delta\Phi_0) \tag{8}$$

with coefficient δ sensitive to the fragrance ingredients properties and having values in a large range from 0 to 175, following the order eugenol > lilial > linalol > phenethyl alcohol > hexyl salicylate > limonene.

Therefore, the influence of solubilization on solution viscosity in spherical micelles for all of the fragrance ingredients studied is weaker than that on the viscosity of the cylindrical micellar solutions. In addition, the strong effect in the cylindrical micelles is observed mainly for surface-active molecules. Eugenol is the fragrance molecule with highest impact on the viscosity of the solutions, whereas limonene is the molecule with the lowest impact.

Reduction of the viscosity (not related to reduction in turbidity) was observed only with surface-active ingredients (phenethyl alcohol and lilial) at very high fragrance ingredient concentration. This finding is related to the melting of the lamellar phase and the transition to micelles. Eugenol (at 2%wt) and linalol (at 3%wt) induce the transition of cylindrical micelles to the hexagonal phase, and the viscosity maintains its high value.

4. Conclusion

The study of the maximum solubilization of various fragrance molecules covering the entire range of hydrophobicity into aqueous micellar solutions containing anionic surfactant revealed a preference of the perfumery ingredients for a given surfactant aggregate's geometry. Correlation between the interfacial affinity of the fragrance molecules and the geometric shape of the surfactant aggregates appropriate for their solubilization is reported.

The perfume ingredients having a micellar core affinity, $EACN_{mix} > 5.5$, are better solubilized in the cylindrical micelles because the solubilization of these fragrance molecules is reduced to a hydrophobic core volume. Phase transition of the surfactant structure is not induced by the solubilization of the highly hydrophobic oils, and thus, the viscosity of the surfactant solution is not modified or slightly increased. Once the MAC of the micellar system is achieved, cloudiness indicates the coexistence of two phases: microemulsion in equilibrium with oil.

In contrast, the perfume ingredients having an interfacial solubility, $EACN_{mix} < 5.5$, are better solubilized in spherical micelles due to their ability to strongly affect the interfacial curvature. The incorporation of these types of fragrance molecules modifies the intermolecular interactions between the head groups of the surfactants themselves and the guest oil. Consequently, the dynamic surfactant-perfume aggregates tend to turn from spherical to cylindrical micelles to liquid crystalline phases. The mixed chemical functionalities of the fragranced molecules do not allow their classification simply into groups of alkanes, alcohols, and esters: the effects of the transition of rod-like to spherical micelles accompanied by a decrease in the viscosity observed by Ulbricht et al. with alcohols, esters, and alkanes were not observed. Moreover, the three molecules considered as alcohols, phenethyl alcohol, eugenol, and linalol, revealed very different behavior when solubilized in spherical and cylindrical micelles; the strongest interfacial influence was observed with eugenol. Phenylethyl alcohol expressed a nontypical alcohol behavior in the cylindrical micelles with a transition from lamellar to micellar phase at very high fragrance ingredient concentration.

Acknowledgements

The author would like to thank to Virginie Soulié, who performed important parts of the experiments during her training period at Firmenich SA.

Conflict of interest

The authors declare no conflict of interest.

Author details

Vera Tchakalova
Formulation and Materials Science Department, DSM-Firmenich SA, S&R, Satigny, Switzerland

*Address all correspondence to: vera.tchakalova@dsm-firmenich.com

References

[1] Tokuoka Y, Uchiyama H, Abe M, Ogino K. Solubilisation of synthetic perfumes by nonionic surfactants. Journal of Colloid and Interface Science. 1992;**152**:402-409

[2] Abe M, Mizuguchi K, Kondo Y, Ogino K, Uchiyama H, Scamehory JF, et al. Solubilisation of perfume compounds by pure and mixtures of surfactants. Journal of Colloid and Interface Science. 1993; **160**:16-23

[3] Tokuoka Y, Uchiyama H, Abe M. Solubilisation of some synthetic perfumes by anionic-nonionic mixed surfactant systems. 2. The Journal of Physical Chemistry. 1994;**98**:6167-6171

[4] Tokuoka Y, Uchiyama H, Abe M, Christian SD. Solubilisation of some synthetic perfumes by anionic-nonionic mixed surfactant systems. 1. Langmuir. 1995;**11**:725-729

[5] Labows JN, Brahms JC, Cagan RH. Solubilization of fragrances by surfactants, Surfactants in Cosmetics: Second Edition, Revised and Expanded, Surfactant Science Series. Vol. 68. 1997. pp. 605-620

[6] Edris AE, Abd El-Galeel M. Solubilisation of some flavor and fragrance oils in surfactant/water system. World Applied Science Journal. 2010;**8**:86-91

[7] Qu Q, Tucker E, Christian SD. Solubilisation of synthetic perfumes by nonionic surfactants and by sulfoalkyl ether ⊠-CDs. Journal of Inclusion Phenomena and Macrocyclic Chemistry. 2003;**45**:83-89

[8] Penfold J, Tucker I, Green A, Jones C, Ford G, Roberts C, et al. Impact of model perfumes on surfactant and mixed surfactant self-assembly. Langmuir. 2008;**24**:12209-12220

[9] Vona SA, Friberg SE, Brin AJ. Location of fragrance molecules within lamellar liquid crystals. Colloids and Surfaces A: Physicochemical and Engineering Aspects. 1998;**137**:79-89

[10] Suratkar V, Mahapatra S. Solubilisation site of organic perfume molecules in sodium dodecyl sulfate micelles: New insights from proton NMR studies. Journal of Colloid and Interface Science. 2000;**225**:32-38

[11] Kayali I, Khan A, Lindman B. Solubilisation and location of phenethyl alcohol, benzaldehyde, and limonene in lamellar liquid crystal formed with block copolymer and water. Journal of Colloid and Interface Science. 2006;**297**:792-796

[12] Kayali I, Qamhieh K, Lindman B. Effect of type of fragrance compounds on their location in hexagonal liquid crystal. Journal of Dispersion Science and Technology. 2006;**27**:1151-1155

[13] Fischer E, Fieber W, Navarro C, Sommer H, Benczédi D, Velazco MI, et al. Partitioning and localization of fragrances in surfactant mixed micelles. Journal of Surfactants and Detergents. 2009;**12**:73-84

[14] Munden R. Effect of perfumes on the viscosity of surfactant systems. Cosmetics & Toiletries. 1988;**103**:65-67

[15] Kanei N, Tamura Y, Kunieda H. Effect of types of perfume compounds on the hydrophilie-lipophile balance temperature. Journal of Colloid and Interface Science. 1999;**218**:13-22

[16] Alam MM, Matsumoto Y, Aramaki K. Effects of surfactant hydrophilicity on

the oil solubilisation and rheological behavior of a nonionic hexagonal phase. Journal of Surfactants and Detergents. 2014;**17**:19-25

[17] Tchakalova V, Fieber W. Classification of fragrances and fragrance mixtures based on interfacial solubilisation. Journal of Surfactants and Detergents. 2012;**15**:167-177

[18] Hoffmann H, Ulbricht W. Transition of rod-like to globular micelles by the solubilisation of additives. Journal of Colloid and Interface Science. 1989;**129** (2):388-405

[19] Christov NC, Denkov ND, Kralchevsky PA, Ananthapadmanabhan KP, Lips A. Synergistic sphere-to-rod micelle transition in mixed solutions of sodium dodecyl sulfate and cocoamidopropyl betain. Langmuir. 2004;**20**:565-571

[20] Tchakalova V, Zemb T, Benczédi D. Evaporation triggered self-assembly in aqueous fragrance-ethanol mixtures and its impact on fragrance performance. Colloids and Surfaces A: Physicochemical and Engineering Aspects. 2014;**460**:414-421

[21] Zoller U, editor. Handbook of Detergents. Part E: Applications. Boca Raton: CRC Press, Taylor & Francis Group; 2008

[22] Rodriguez C, Shresta LK, Varade D, Aramaki K, Maestro A, Lopez-Quentala A, et al. Formation and properties of reverse micellar cubic liquid crystals and derived emulsions. Langmuir. 2007;**23**: 11007-11014

[23] Montalvo G, Valiente M, Rodenas E. Rheological properties of the L phase and the hexagonal, lamellar, and cubic liquid crystals on the CTAB/benzyl alcohol/ water system. Langmuir. 1996;**12**:5202-5208

[24] Ansmann A, Busch P, Hensen H, Hill K, Krächter HU, Müller M. Personal care formulations. In: Showell MS, editor. Handbook of Detergents, Part D: Formulation. Boca Raton: CRC Press; 2006. pp. 207-260

[25] Abdel-Rahem R, Regerr M, Hlucha M, Hoffman H. Rheology of aqueous solutions containing SLES, CAPB and microemulsion: Influence of co-surfactant and salt. Journal of Dispersion Science and Technology. 2014;**35**:64-75

[26] Christov NC, Denkov ND, Kralchevsky PA, Broze G, Mehreteab A. Langmuir. 2002;**18**:7880-7886

[27] Miller CA. Solubilisation in surfactant systems. In: Birdi KS, editor. Handbook of Surface and Colloid Chemistry. Boca Raton: CRC Press; 2009. pp. 415-438

[28] Todorov et al. Journal of Colloid and Interface Science. 2002;**24**:371-382

[29] Tchakalova V, Testard F, Wong K, Parker A, Benczedi D, Zemb T. Solubilisation and interfacial curvature in microemulsions: I. Interfacial expansion and co-extraction of oil. Colloids and Surfaces A: Physicochemical and Engineering Aspects. 2008;**331**:31-39

[30] Saito Y, Abe M, Sato T. Effects of micellar structure on solubilisation of n-octane and n-octanol. Colloid and Polymer Science. 1993;**271**:774-779

[31] Israelachvili J. The science and applications of emulsions – An overview. Colloids and Surfaces A: Physicochemical and Engineering Aspects. 1994;**91**:1-8

[32] Evilevitch A, Olsson U, Jonsson B, Wennerstrom H. Langmuir. 2000;**16**: 8755-8762

Section 2

Microemulsions and Nanoemulsions

Chapter 4

Nano Emulsions: A Novel Targeted Delivery of Cancer Therapeutics

Gudikandula Krishna and Dasari Thrimothi

Abstract

The primary focus in the management of cancers is the administration of therapeutic medicines to the specific tumor locations. The therapy of cancer is significantly hindered by the presence of medicines with high lipophilicity, limited absorption, and non-specific toxicity. Nano emulsions are a type of heterogeneous particle system that has demonstrated its reliability in the realm of nanotechnology. Oil in water nano emulsions possess the potential to significantly transform traditional cancer treatment due to their hydrophobic core, small size, and exceptional durability. Nano emulsions has notable characteristics such as the ability to encapsulate medications with low solubility, the capacity to selectively target tumor cells, and the ability to overcome multidrug resistance (MDR). These traits have demonstrated significant advantages in the treatment of several cancer types. Currently, researchers are conducting experiments and studies to investigate the efficacy and safety of several nano emulsion formulations in the therapeutic treatment of malignancies. The purpose of this chapter is to provide an overview of the current uses of nano emulsions in the specific delivery of anticancer drugs.

Keywords: nanotechnology, nano emulsions, cancer therapy, targeted delivery, formulation

1. Introduction

The primary challenge in cancer diagnosis and therapy lies in achieving precise and effective drug delivery to cancer cells while minimizing their harmful impact on adjacent healthy cells. Chemotherapy medications exert their effects by selectively killing rapidly dividing cells, including hair follicles, red blood cells, bone marrow, lymphatic cells and gut epithelia. Consequently, these drugs are not suitable for prolonged treatment. The lack of specificity of chemotherapy medications can result in fatal harm to nearby healthy cells that are actively dividing [1]. In contrast, many extremely promising medications are eliminated from the development process because they have low solubility in water [2]. Therefore, creating an effective delivery system that can specifically target cancer cells and transport therapeutic and imaging chemicals to tumors is imperative. This will enhance the chances of early-stage diagnosis and successful treatment. Nano systems such as nanoparticles, liposomes, micelles, nano capsules, and nano emulsions have emerged as nanocarriers for delivering hydrophobic medicines to specific locations.

Nano emulsions are a mixture of two liquids that do not mix well (such as oil and water), with an average droplet size ranging from 50 to 200 nanometers [3]. Nano emulsions, called nanosized micelles or micellar nanoparticles, are stabilized using surfactants, known as emulsifying agents. Emulsifying agents are molecules that have both hydrophilic and hydrophobic properties. They can lower the tension between two liquid phases that do not mix, such as oil and water. This is achieved by selectively attaching to the interfaces between these phases. Emulsifying agents possess a polar head group that selectively aligns with polar liquids like water and a nonpolar hydrocarbon tail that prefers nonpolar liquids like oil. They exhibit transparency or translucency and possess a substantial surface area due to their diminutive particle size. Nano emulsions are often formulated using excipients recognized by the United States Food and Drug Administration (FDA) as Generally Recognized as Safe (GRAS) grade. Nano emulsions offer a significant benefit by enabling the encapsulation of hydrophobic pharmaceuticals within their hydrophobic core. This results in enhanced drug loading and delivery capabilities. The solubility of the poorly water-soluble medication resveratrol was enhanced by the nano emulsion, including ethyl oleate, Tween 80, and PEG400, through increased drug loading within the nano emulsion. Nano emulsion effectively shields the medication from enzymatic breakdown and hydrolysis by enclosing it in the core. Additionally, it hinders the identification of the drug by macrophages, which are part of the mononuclear phagocytic system (MPS), resulting in an extended circulation period. The extended circulation duration enables the Nano emulsions to aggregate more extensively in regions with compromised and permeable blood vessels, such as tumors, by leveraging the enhanced permeability and retention (EPR) phenomenon [4].

2. The cancer burden is increasing

The prevalence of cancer is increasing globally. The global incidence of cancer was 14 million in 2012, resulting in 8 million deaths. However, it is projected that these figures will rise to 22 million new cases and 13 million deaths per year by 2030 [5]. Changes in the economy, society, and way of life are the causes of the increase as well as aging populations. Cancer and human development are strongly correlated with socioeconomic status (**Figure 1**) [5].

The worldwide prevalence of cancer is not increasing uniformly [5, 6]. Due to the uneven rates of globalization in economics and lifestyle practices, there are disparities in the exposure to causal factors during transitions. Cancer is the primary cause of mortality in numerous affluent nations and is projected to become the top cause in middle- and low-income nations shortly. Currently, cancer is more prevalent in high-income areas, whereas mortality rates are higher in middle- and low-income areas. The majority of the projected rise in cancer prevalence is anticipated to take place in middle- and low-income nations, which are the least equipped to address this issue.

A significant challenge in many cancer treatments is the effective delivery of medicines to the tumor site while minimizing adverse effects. The effectiveness of therapies throughout the body has relied on their physical and chemical characteristics, such as size, ability to diffuse, and how well they bind to plasma proteins. However, tumors have a complex and diverse network of blood vessels that tend to push medications away from them. This impedes the precise administration of drugs to the specific location of the tumor. In addition, the traditional methods of administering anticancer treatments pose several challenges, including the emergence of

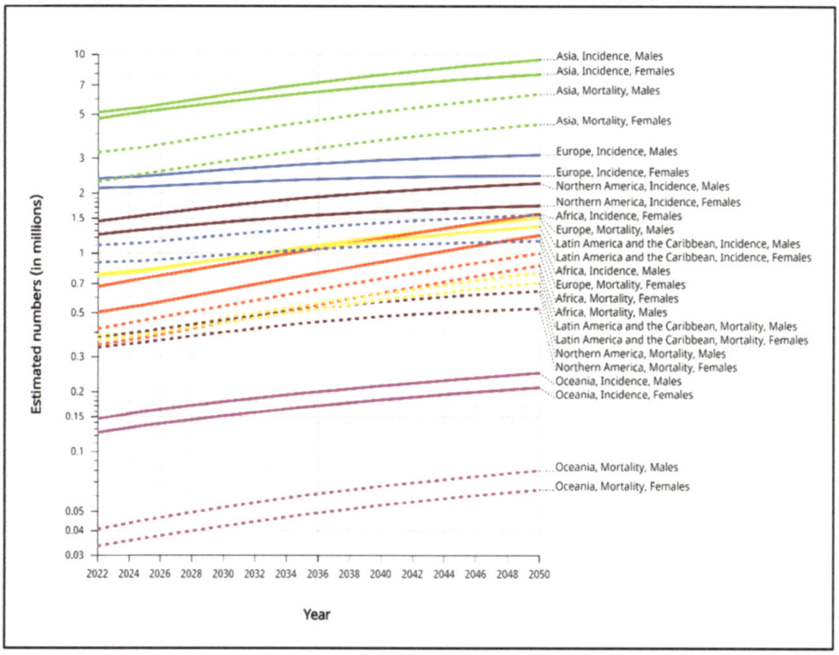

Figure 1.
Represents the global incidence prediction of cancer from 2022–2050. Source: World Health Organization (cancer tomorrow).

multidrug resistance (MDR) in cancer cells, significant toxicity to non-tumor cells, and low specificity [7].

Furthermore, most medicines designed for cancer treatment have a significant degree of hydrophobicity. Nanocarriers have developed as viable drug delivery carriers to address the challenge of incorporating hydrophobic drugs and delivering them to specific tumor locations. The utilization of nanocarriers has experienced a substantial rise in recent years due to their multiple favorable qualities in medicine delivery. Drug delivery mechanisms are called nanocarriers, typically consisting of colloids with a smaller particle size than 500 nanometers [8]. Nanocarriers including nanoparticles, dendrimers, liposomes, Nano emulsions and nano capsules can effectively utilize the tumor microenvironment to enhance the therapeutic efficacy of the anticancer medicine they carry [9].

Nano emulsions are a type of colloidal drug carrier system consisting of submicron droplets, typically ranging from 10 to 1000 nm [10]. They are one of the various nanocarrier systems available. Nano emulsions are formed of two immiscible liquids and have a spherical nanoscale structure. They can exist in three forms: oil-in-water (O/W), water-in-oil (W/O), or bi-continuous. Nano emulsions provide numerous favorable characteristics that make them suitable for application as innovative drug carrier systems. Nano emulsions are tiny, enabling them to easily traverse the highly vascularized tumor environment and accumulate within it.

Furthermore, their spherical shape grants them a substantial surface area that can enhance adsorption capabilities. Moreover, the O/W Nano emulsions include a lipophilic core that effectively encapsulates medicines with high hydrophobicity. Furthermore, these substances are harmless, resistant to change, non-irritating, capable of significantly enhancing the drug's ability to be absorbed by the body, and suitable for precise targeting at specific locations [11]. Primarily, due to the sustained and regulated medication release pattern, it is possible to reduce the dosage and frequency of injections over the entire course of treatment [12]. Nano emulsions have unique characteristics that make them an innovative vehicle for the precise and regulated administration of cancer treatments. The present chapter focuses on developing Nano emulsions for precise and controlled administration of anti-cancer drugs, explicitly targeting cancer cells. Additionally, it explores the potential applications of these Nano emulsions in treating different cancer types. Finally, the chapter provides a comprehensive overview of current and finished clinical trials investigating the practical application of nano emulsion formulations with efficacy and safety.

3. Nano emulsions – A brief overview

Recently, Nano emulsions have been employed for targeted drug delivery to specific locations. For more than ten years, researchers have investigated the potential uses of Nano emulsions in various methods of drug administration, including oral [7], parenteral [8], transdermal [7], ophthalmic [9], pulmonary [10], nasal medication [11] and nasal vaccine delivery [12]. Furthermore, researchers have investigated the potential of utilizing nano emulsion as a carrier for delivering a diverse range of drugs, including parenteral nutrition, analgesics, anesthetics, steroids, chemotherapeutics and tumor vaccines. Here are a few examples of commercially available nano emulsion products. Soybean oil, egg yolk phospholipid, and glycerin-based emulsions have been extensively utilized in healthcare settings for parenteral nutrition, such as in the products Intralipid® and Vitalipid®. These lipids offer a plentiful supply of crucial fatty acids, including omega-3 and omega-6 fatty acids, linoleic acid, calories not derived from glucose, and vitamins E and K. In addition, Nano emulsions have been extensively utilized as colloidal carriers for the precise administration of various chemotherapeutic medications and diagnostic and imaging agents.

Furthermore, nanoparticles can be customized with precise ligands to target cells, tissues, or organs selectively in conjunction with a fluorescent dye for imaging

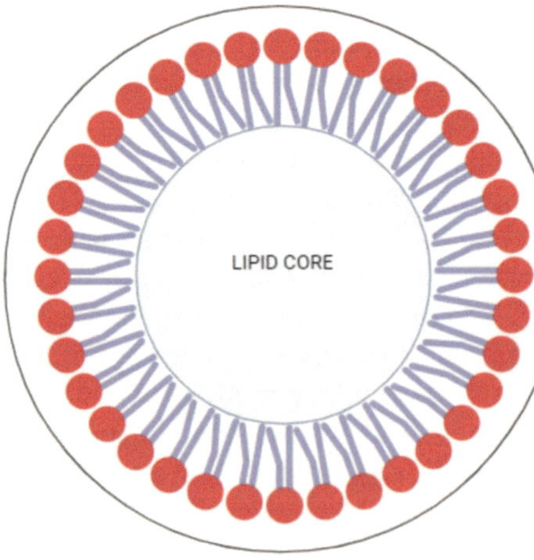

Figure 2.
Represents the internal structure of oil-in-water nano emulsion's.

purposes. The extensive investigation of these versatile nanoparticles is focused on their use in cancer treatment. Nanoparticles are linked with versatile components to facilitate drug detection and targeted administration for cancer treatment (**Figure 2**) [13].

4. Nano emulsion and anticancer therapy

Cancer is a group of diseases that include the abnormal and uncontrolled development of cells [14]. Cancer arises from molecular-level changes in genes, where various subsets of genes experience genetic modifications, such as the activation of oncogenes or the inactivation of tumor suppressor genes. Consequently, there is an uncontrolled growth of cancer cells, infiltration of tissues, and impairment of organ function [15]. Tumor tissues exhibit robust angiogenesis and a high density of blood vessels to support their growth. However, they also have a flawed vascular structure and limited lymphatic drainage, which collectively contribute to the phenomenon known as the enhanced permeation and retention (EPR) effect [16, 17]. Tumor genes frequently exhibit genetic change during their development rather than remaining constant. The inherent intricacy of the tumor microenvironment and the presence of P-glycoprotein (Pgp) typically serve as obstacles to conventional chemotherapy by impeding the drug's access to the tumor mass. Delivery of therapeutic medications in vivo encounters many physiological barriers, such as hepatic and renal clearance, enzymolysis and hydrolysis, and endosomal/lysosomal degradation (**Figure 3**) [18, 19].

Several chemicals are being investigated for their potential use in medication delivery, particularly cancer therapeutic technology. Formulating pharmaceutical items with optimal therapeutic efficacy and minimal or insignificant adverse effects is beneficial. An inherent drawback of anticancer medications is their limited specificity

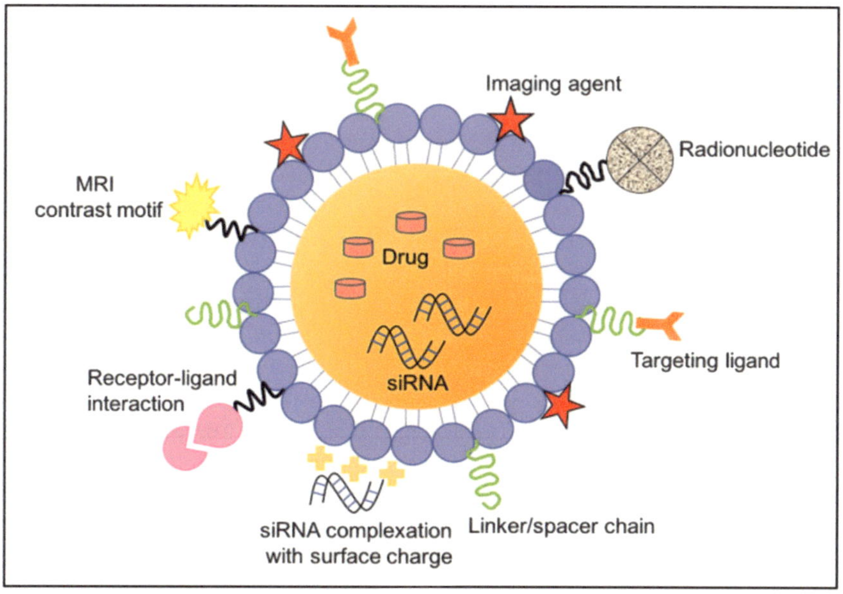

Figure 3.
Represents the a visual of a multi-purpose nano emulsion system.

for tumor tissue, leading to significant adverse effects and diminished success rates in achieving a complete cure. Hence, the typical medication delivery system has significant challenges in effectively targeting the aberrant cells [14]. Furthermore, the effectiveness of anticancer medications is constrained by their inadequate characteristics, including low solubility, a limited therapeutic range, and excessive toxicity towards healthy tissues. These factors may contribute to the failure of cancer treatment [20, 21]. Nano emulsions can transport a concentrated number of chemotherapeutic medications to malignant tissues without causing harm to cells and organs in the body's overall circulation. They have demonstrated significant efficacy in medication delivery for cancer treatment. Nano emulsions can be easily detected using a range of imaging modalities, their application in therapeutic targeting and their capacity to enhance drug bioavailability. The incorporation of lipophilic anticancer medicines into oil-in-water Nano emulsions has demonstrated significant benefits. The oil phase of the emulsion functions as a solubilizer for the lipophilic chemical, increasing the drug's solubility in an emulsion system.

Consequently, a smaller quantity of the medication must be provided in contrast to a solution including water. The discomfort caused by intravenously injected medications can be reduced by using a lipid emulsion. This method involves exposing the tissue to a lower drug concentration and eliminating the use of a carrier that irritates the tissue [22].

5. Different types of cancer and their nano emulsions

5.1 Nano emulsions for the treatment of cancer

Nanotechnology has demonstrated its efficacy as a viable approach for cancer therapy, prompting researchers to concentrate on treating various cancer forms. The subsequent part provides a concise overview of the latest breakthroughs in the prevalent types of cancer (onco).

5.2 Nano emulsion for treatment of colon cancer

Colon cancer accounts for a significant proportion of global cancer mortality [23]. Under this group, the sub-classification comprises hereditary non-polyposis colorectal cancer, familial adenomatous polyposis, colitis-associated cancer, and spontaneous colon cancer [24]. Surgical intervention and the administration of medicinal plants, immunotherapy, radiation, and chemotherapy are commonly employed in managing colon cancer. However, the survival rate declines around five years post-surgery as a result of recurrence and metastasis, indicating that the primary cause of mortality is not the tumor itself [25]. Cancer migration and invasion are facilitated by epithelial-mesenchymal transition (EMT), a process in which epithelial cells transform into mesenchymal cells, resulting in alterations in cell structure and enhanced adhesion and migratory capabilities [26].

Tomatoes contain lycopene (LP), which has several valuable properties, including protection against chronic illnesses, anti-leukemia proliferation activity and cells associated with colon cancer, and the ability to further induce cell cycle arrest on some tumor cells [27]. If LP's low stability and bioavailability were not a problem, its mechanism might be used in cancer therapy [28]. The study's goal was to solve the given challenge by creating a nano emulsion using LP. Gold nanoparticles (AN) are also encapsulated in the nano emulsion formulation. In addition to its role as a drug carrier, AN can be combined with cell receptor ligands supplied for particular cell targeting [29]. However, because AN promotes the migration of human fibroblast cells, excessive quantities of the substance can be hazardous [30]. Polymeric materials, liposomes, or other lipid-based assemblies, including LP-derived chemicals, can be added to lessen this effect [31].

The emulsifier in this particular formulation is Tween 80®, an aqueous AN solution in the water phase and the oil phase is oil with LP. The human colon cancer line HT-29 can utilize it. The effectiveness of the formulation is demonstrated by the study of the impact on the cell line of AN alone, LP alone, and AN and LP combined [23].

The toxicity of AN is influenced by its size; the greater its impact on HT-29 cells, the smaller the AN. The number of early apoptotic cells may rise with an increase in the volume of LP incorporated. The number of late apoptotic, early apoptotic, and necrotic cells increases with combined therapy and the Nano emulsions. However, the application of Nano emulsions causes necrotic and apoptotic cells. Only the stability of the Nano emulsions is enhanced by the emulsifier; it has no appreciable effect on cells. Low AN and LP dose-treated Nano emulsions cause cells to undergo apoptosis and increase the expression of PARP-1 and Bax while reducing procaspases 3 and 8 as well as Bcl-2 expression, tumor indicators [23].

5.3 Nano emulsions for treatment of ovarian cancer

Chemotherapeutics containing platinum (Pt) are employed in various cancer therapies. Carboplatin and cisplatin, which both have platinum (Pt) in their composition, exhibit superior efficacy in improving survival rates compared to all other treatments for ovarian cancer [32]. DNA structure is broken by PT chemicals that create intra- and inter-strand cross-connections [33]. As Pt activity ultimately results in the death of healthy cells, it also hurts cells other than cancerous ones. Additionally, cancer cells can establish defense mechanisms against Pt, such as triggering enzymes or DNA repair pathways or encouraging the existence of membrane pumps. Treatments for ovarian cancer, therefore, always take this into account, as does the tumor's sensitivity to platinum [34].

Formulations to address issues with toxicity and resistance have been developed thanks to the evolution of nanomedicine. However, the procedure is complicated because of the characteristics of Pt, especially its lipophilicity [35]. Nano emulsions can enhance pt-related drug delivery and efficiency by combining large hydrophobic medicines with particular ligands to enable targeted delivery. This study created a Nano emulsions to contain myrisplastin, a new platinum Pt-based medication with the pro-apoptotic compound C6-ceramide [36].

Additionally, an EGFR-binding peptide, a surface ligand, and an imaging agent called gadolinium were added to the Nano emulsions. The purpose of this work was to examine the effects of this Nano emulsions on ovarian cancer cells, namely SKOV3, A2780, and A2780CP [34].

Epidermal growth factor receptor (EGFR)-expressing SKOV3 cells were resistant to cisplatin, with an inhibitory concentration (IC50) of 18 μM, according to a cytotoxicity screening. By encapsulating myrisplatin instead, cytotoxicity rose considerably, both in Nano emulsions with and without targets. The toxicity of the targeted Nano emulsions is double that of the non-targeted formulations. When ceramide is added to the encapsulation, the cytotoxicity changes significantly, indicating that the two compounds work in concert. Cisplastin is 50.5 times less effective than the intended

Nano emulsions of myrisplatin and ceramide. Myrisplatin caused higher harmful effects in A2780 and A2780CP (not expressing EGFR) than cisplatin [34].

Zeng et al. created vitamin E Nano emulsions with paclitaxel that might block P-gp transport and change the expression of Bac & BCL-2, two factors linked to tumor drug resistance. The human ovarian cancer cell line A2780, resistant to paclitaxel, was used to evaluate their impact. Taxol reduced the potential of mitochondria and increased the antiproliferation action. According to the authors, vitamin E-derived multifunctional Nano emulsions combined with anticancer medications may effectively treat cancer multidrug resistance [37].

5.4 Nano emulsions therapy for prostate cancer

Prostate cancer (PrC) has been the cause of an increase in fatalities over the past ten years, with 70% of treated patients experiencing a recurrence and eventual transition to an incurable state [38]. Tumor-initiating cells (TICs), also known as cancer stem cells (CSCs), are the primary cause of cancer metastasis and treatment resistance [39]. Research indicates that cancer cells that exhibit CSC markers—particularly CD133 and CD4—not only show a correlation with medication resistance but also continue to multiply following treatment [40]. Drug resistance in CSCs may result from increased drug efflux transporter expression, anti-apoptotic pathway activation, apoptotic mechanism deactivation, and enhanced DNA damage and repair process efficiency [41].

Prostate cancer therapy has an issue since it targets populations of rapidly proliferating cancer cells rather than subpopulations like CSCs. Additionally, cell lines with high passage numbers are used in preclinical investigations to assess anti-cancer medicines in developing drugs to treat prostate cancer. The genomic and epigenomic characteristics these cell lines adopt have little to nothing in common with the initial tumor [42]. The cell line (PPT2) used in this work is obtained from a patient with prostate cancer with a deficient passage number, preserving the cell's immaturity and stem-like characteristics. Because PPT2 cells have genes linked to anti-apoptotic signaling and drug resistance, they are an ideal model for research on CSC-targeted therapy [43].

A commonly employed medicine in the treatment of prostate cancer is Abraxane®. A paclitaxel pro-drug called Abraxane® was created to improve its solubility using a formulation of human serum albumin-bound nanoparticles. However, issues with MDR cancer cells have been observed with paclitaxel [41]. Drug resistance can be effectively combated with SBT-1214, a new-generation taxoid. Docosahexaenoic acid (DHA), a naturally occurring polyunsaturated fatty acid (PUFA) with a strong affinity for its primary bloodstream transporter (human serum albumin), can be conjugated with this chemical to assist target cancerous cells. The combination of DHA and paclitaxel led to a slight reduction in ABC and P-gp transporters [44]. The Nano emulsions formulation, named DHA-SBT-1214, created in this research incorporates phospholipids and fish oil. The medication's affinity for fish oil will enhance drug encapsulation. The Nano emulsions is hypothesized to target the CSC-initiated PPT2 cell line by utilizing the EPR effect, ultimately leading to the death of cancer cells [41].

Drug development targeting cells specific for tumor initiation may benefit from using patient-derived CSC-enriched PPT2 cells. By combining DHA with SBT-1214, the formulation can remain in the bloodstream for longer. Incorporating the hydrophobic medication into a Nano emulsions formulation leads to efficient drug delivery.

Consequently, adding PEG to the surface modifies it and prolongs the duration of drug circulation, thereby promoting accumulation due to the enhanced permeability and retention (EPR) effect. The effective cellular internalization indicates that the Nano emulsions formulation can deliver its payload with greater efficiency than the drug solution [41].

5.5 Leukemia through nano emulsions

Cancer is a prominent contributor to global mortality, with leukemia being the primary cause of cancer-related fatalities in children. Nano emulsions have been employed as biocompatible systems in this particular scenario to encapsulate pharmaceuticals and enhance therapeutic outcomes by reducing hazardous side effects.

Various researchers have employed lipid Nano emulsions as a practical approach for encapsulating drugs in the context of cancer therapy. Moura and colleagues produced lipid Nano emulsions that could attach to LDL receptors to concentrate chemotherapeutic drugs in tissues like tumor tissues that overexpress the low-density lipoprotein receptor. Because the uptake of the Nano emulsions is substantially higher than that of the free drug, boosting the toxicity against tumoral cells, the authors encapsulate methotrexate for leukemia treatment and evaluate them in vitro [45].

Winter et al. synthesize and evaluate chalcone-encapsulating Nano emulsions for leukemia both in vitro and in vivo. The authors show similar anti-leukemic effects for the free chalcones and the Nano emulsions, demonstrating that the Nano emulsions produced cause cancer cells to undergo apoptosis in vitro. On the other hand, chalcones-loaded Nano emulsions were less hazardous to VERO cells than free chalcones. In vivo observations revealed similar outcomes. Oxidative stress and weight gain were observed to be reduced by free chalcones, along with an inflammatory response [46].

5.6 Breast cancer: nano emulsions therapy

Worldwide, breast cancer represents 23% of all newly identified cases in women and 13.7% of cancer-related deaths. The scarcity of chemotherapeutic medications is the disparity in their concentration between tumors and other organs, leading to higher toxicity levels due to their lower accumulation in tumors. Several strategies have been developed to improve the treatment of this ailment in patients.

A viable treatment for breast cancer may involve using natural substances, such as the *Nigella sativa* L. essential oil used in the Nano emulsions created by Periasamy and associates [47]. By causing apoptosis in MCF-7 breast cancer cells, this Nano emulsions has anti-cancer capabilities in vitro. For breast cancer treatment, this Nano emulsions may help encase active medications [47].

The technique of utilizing local administration and the invention of C6 ceramide Nano emulsions has been employed to treat breast cancer. The authors want to specifically target malignancies and pre-tumor lesions in a localized manner while minimizing the adverse effects on the entire body by utilizing both local administration and medication distribution by Nano emulsions.

The researchers created Nano emulsions containing ceramide and enhanced their sticky properties by modifying the surface with chitosan. The concentration of C6 ceramide required to reduce the viability of MCF-7 cells by 50% (EC50) was reduced by a factor of 4.5 when it was nanoencapsulated compared to when it was in solution. Additionally, when tributyrin, a pro-drug of butyric acid, was included in the oil phase of the Nano emulsions, a further decrease of 2.6-fold was seen.

When administered intraductally, the Nano emulsions resulted in a more extended localization of the medication in the mammary tissue for over 120 hours compared to its solution [48]. Natesan and his colleagues utilized chitosan to manufacture Nano emulsions [49]. The researchers enclosed camptothecin within Nano emulsions and evaluated their performance both in laboratory settings (in vitro) and in living organisms (in vivo), demonstrating the superior effectiveness of the formulations in comparison to the unbound drug [49].

5.7 Melanoma nano emulsions therapy

Melanoma, the most lethal type of skin cancer, accounts for about 80% of deaths associated with skin cancer. The primary issue related to the management of melanoma is the limited effectiveness of current treatment methods, primarily due to the inadequate response of chemotherapeutic drugs and the inherent resilience of melanoma cells. Conventional therapies for advanced melanoma and spreading melanoma typically yield unsatisfactory outcomes, resulting in severe adverse reactions and low overall survival rates. Consequently, researchers are investigating more recent combinations of anti-melanoma medications and advanced approaches, including nanotechnology, such as Nano emulsions.

Paclitaxel-encapsulating lipid Nano emulsions that bind to low-density lipoprotein (LDL) receptors were created by Kretxer and colleagues. This reduces drug-associated toxicity and increases the antitumoral effect. Additionally, simvastatin association was evaluated in mice with melanoma, showing that this medication enhanced antitumoral activity in conjunction with paclitaxel Nano emulsions but not with free paclitaxel. This could be attributed to the fact that statins strengthen the expression of LDL receptors, which are responsible for the internalization of lipid Nano emulsions [50]. Several other authors incorporated cholesterol derivatives, specifically 7-ketocholesterol, into lipid core Nano emulsions and evaluated their effects in vivo using a murine melanoma cell line. The results demonstrated that the Nano emulsions significantly reduced tumor size by over 50%, increased the necrotic area, and decreased the vasculature within the tumor. The cellular uptake of the cholesterol Nano emulsions by tumor cells was facilitated via the LDL receptor, indicating a process of internalization. Notably, a single administration of these Nano emulsions resulted in the death of 10% of melanoma cells [51].

Monge-Fuentes and colleagues employed an alternative method, specifically photodynamic treatment, utilizing acai oil in Nano emulsions. The Nano emulsions was used as a photosensitizer both *in vitro* and *in vivo*. Treatment was applied to B16F10 melanoma cell lines and NIH/3 T3 normal cells, resulting in a significant 85% cell death, specifically in the melanoma cells. However, the normal cells maintained a high level of viability. C57BL/6 mice with tumors administered acai oil Nano emulsions exhibited a significant reduction in tumor volume by 82% [52].

5.8 Nano emulsions therapy for lung cancer

Paclitaxel (PTX) is a chemotherapeutic agent frequently employed in treating lung, breast, ovarian malignancies and pancreatic. It can disrupt the disintegration of microtubules during cellular division resulting in mitotic arrest, apoptosis and suppression of cell activities [53]. PTX exhibits a minimal ability to dissolve in water, necessitating the development of other formulations, such as Taxol®, which incorporates Cremophor-EL® and ethanol. Nevertheless, Cremophor-EL® is recognized for its hazardous properties, necessitating the exploration of targeted compounds for this formulation [54].

The potential of hyaluronic acid (HA) for effectively delivering PTX to cancer cells has been studied. It has a negative charge and selectively attaches to a cluster of differentiation 44 (CD44), a tumor cell marker that is abundantly expressed [55]. This study aims to assess the effectiveness of a Nano emulsions carrier carrying HA and PTX, namely the HA-complexed PTX Nano emulsions (HPNs), in targeting tumors that express CD44 in a non-small lung carcinoma cell line (NCI-H460) [56].

With a low polydispersity index indicating formulation homogeneity, a size that permits a long half-life and zeta potential suitable to stabilize the formulation, and the intended spherical morphology, HPN showed excellent physical–chemical particle properties. Tumor weight evaluation revealed that HPN and PTX Nano emulsions both slowed the growth of tumors but that HPN's targeted component, HA, improved the effectiveness of treatment. The body weight results indicate no significant alterations observed in the groups who received PTX Nano emulsions and HPN. This finding confirms that these treatments are less harmful to healthy tissues [56].

Chang et al., investigate the anti-cancer potential of Curcuma longa Linnaeus's curcuminoid preparations. They created Nano emulsions to investigate the inhibitory mechanism linked to the anti-cancer effect against lung cancer cells. The cell cycle progression in lung-cancer cells was delayed at the G2/M phase by both the Nano emulsions and curcuminoid extract therapies, although the specific biological mechanism may vary. H460 cells exhibit a higher vulnerability to apoptosis than A549 cells when exposed to curcuminoids and their applications in Nano emulsions across various cancer cell lines [57].

6. Nanotheragnostics: a use for nano emulsions

Nanotheragnostics is a novel approach that harnesses the capabilities of nanotechnology for the specific goals of diagnostics and imaging. The objective of this technique is to develop nanoscale molecules that can perform both therapeutic and diagnostic activities [58].

Researchers have placed a particular emphasis on the treatment and diagnostics of cancer, and this method is linked to several nanotechnological approaches including Nano emulsions [58]. Recent progress in this domain has facilitated the analysis of individual tumors, forecasting Nano emulsions -tumor interactions and the development of personalized nanomedicines for therapy [58].

In this way, Fernandes et al. created perfluorohexane Nano emulsions as a novel contrast agents and drugs delivery systems for in vivo ultrasound and photoacoustic cancer imaging. These Nano emulsions provide deeper tissue penetration and higher spatial resolution than traditional optical methods. This neural enhancement offers a non-intrusive option for cancer imaging and therapy in patients [59].

A similar approach was taken by Wu et al., who created magnetic Nano emulsions hydrogels that cause magnetic tumor regression by heating malignancies to a ferrofluid temperature [60]. In addition, Niravkumar et al., designed Nano emulsions that contained three different acid platins: dipalmiplatin, distearyplatin and dimyrisplatin. The researchers created Nano emulsions containing fatty acids that specifically attach to utilize receptor-mediated endocytosis and the folate receptor α (FR-α) to transport Pt via mechanisms of resistance of the cell surface. FR-α is seen in high levels in various cancerous diseases including ovarian cancer [61]. Roberts et al., employed sonophore molecules to conduct multispectral optoacoustic tomography (MSOT) on malignancies. They merged near-infrared and strongly absorbing dyes incorporated into Nano emulsions, facilitating the non-invasive *in vivo* detection of malignancies via MSOT.

7. Conclusions

Nano emulsions are extensively investigated for their potential to deliver cancer drugs. However, the FDA approves no cancer treatment products for clinical usage. Nevertheless, some Nano emulsions products are presently being examined in clinical trials. The challenges above arise due to quality control, material safety, and scale-up operations. The design and development of multifunctional Nano emulsions at a more significant size, capable of carrying out several activities, pose substantial challenges. Before their clinical application, it is imperative to thoroughly evaluate the in vivo distribution, safety, and metabolism of Nano emulsions and their functional moieties. In addition, the concurrent optimization and evaluation of passively targeted Nano emulsions could provide valuable insights into the impact of design factors on multifunctional Nano emulsions.

Author details

Gudikandula Krishna* and Dasari Thrimothi
Department of Microbiology, Kakatiya University, Warangal, Telangana, India

*Address all correspondence to: krishnagudikandula@gmail.com

References

[1] Mahato R, Tai W, Cheng K. Prodrugs for improving tumor targetability and efficiency. Advanced Drug Delivery Reviews. 2011;**63**:659-670

[2] Sareen S, Mathew G, Joseph L. Improvement in solubility of poor water-soluble drugs by solid dispersion. International Journal of Pharmaceutical Investigation. 2012;**2**:12

[3] Ganta S, Talekar M, Singh A, Coleman TP, Amiji MM. Nano emulsions in translational research—Opportunities and challenges in targeted cancer therapy. AAPS PharmSciTech. 2014;**15**:694-708

[4] Maeda H, Wu J, Sawa T, Matsumura Y, Hori K. Tumor vascular permeability and the EPR effect in macromolecular therapeutics: A review. Journal of Controlled Release. 2000;**65**:271-284

[5] Whiteman DC, Wilson LF. The fractions of cancer attributable to modifiable factors: A global review. Cancer Epidemiology. 2016;**44**:203-221. DOI: 10.1016/j.canep.2016.06.013

[6] Lortet-Tieulent J, Georges D, Bray F, Vaccarella S. Profiling global cancer incidence and mortality by socioeconomic development. International Journal of Cancer. 2020;**147**:3029-3036. DOI: 10.1002/ijc.33114

[7] Rosenblum D, Joshi N, Tao W, Karp JM, Peer D. Progress and challenges towards targeted delivery of cancer therapeutics. Nature Communications. 2018;**9**(1):1410

[8] Din FU, Aman W, Ullah I, Qureshi OS, Mustapha O, Shafique S, et al. Effective use of nanocarriers as drug delivery systems for the treatment of selected tumors. International Journal of Nanomedicine. 2017;**12**:7291-7309

[9] Lôbo GC, Paiva KL, Silva ALG, Simões MM, Radicchi MA, Báo SN. Nanocarriers used in drug delivery to enhance immune system in cancer therapy. Pharmaceutics. 2021;**13**(8):1167

[10] Jaiswal M, Dudhe R, Sharma PK. Nano emulsions: An advanced mode of drug delivery system. 3. Biotech. 2015;**5**:123-127

[11] Kishore S, Kumar SA, Roy T, Bhardwaj A, Malik S. Nano emulsions formulations in targeted delivery of cancer therapeutics. In: Advancements in Controlled Drug Delivery Systems. IGI Global; 2022. pp. 44-72

[12] Adepu S, Ramakrishna S. Controlled drug delivery systems: Current status and future directions. Molecules. 2021;**26**(19):5905

[13] Khandavilli S, Panchagnula R. Nano emulsions as versatile formulations for paclitaxel delivery: Peroral and dermal delivery studies in rats. Journal of Investigative Dermatology. 2007;**127**:154-162

[14] Talekar M, Ganta S, Singh A, Amiji M, Kendall J, Denny WA, et al. Phosphatidylinositol 3-kinase inhibitor (PIK75) containing surface functionalized Nano emulsions for enhanced drug delivery, cytotoxicity and pro-apoptotic activity in ovarian cancer cells. Pharmaceutical Research. 2012;**29**:2874-2886

[15] Gallarate M, Chirio D, Bussano R, Peira E, Battaglia L, Baratta F, et al. Development of O/W Nano emulsions

for ophthalmic administration of timolol. International Journal of Pharmaceutics. 2013;**440**:126-134

[16] Nesamony J, Shah IS, Kalra A, Jung R. Nebulized oil-in-water Nano emulsions mists for pulmonary delivery: Development, physico-chemical characterization and in vitro evaluation. Drug Development and Industrial Pharmacy. 2014;**40**:1253-1263

[17] Mistry A, Stolnik S, Illum L. Nanoparticles for direct nose-to-brain delivery of drugs. International Journal of Pharmaceutics. 2009;**379**:146-157

[18] Bielinska AU, Janczak KW, Landers JJ, Markovitz DM, Montefiori DC, Baker JR Jr. Nasal immunization with a recombinant HIV gp120 and Nano emulsions adjuvant produces Th1 polarized responses and neutralizing antibodies to primary HIV type 1 isolates. AIDS Research and Human Retroviruses. 2008;**24**(2):271-281

[19] Tiwari S, Tan Y-M, Amiji M. Preparation and in vitro characterization of multifunctional Nano emulsions for simultaneous MR imaging and targeted drug delivery. Journal of Biomedical Nanotechnology. 2006;**2**:217-224

[20] Mathur V, Satrawala Y, et al. Solid lipid nanoparticles in cancer therapy, international. Journal of Drug Delivery. 2010;**2**:192-199

[21] Sarkar FH, Banerjee S, Li YW. Pancreatic cancer: Pathogenesis, prevention and treatment. Toxicology and Applied Pharmacology. 2007;**224**(3):326-336

[22] Byrne JD, Betancourt T, Brannon-Peppas L. Active targeting schemes for nanoparticle systems in cancer therapeutics. Advanced Drug Delivery Reviews. 2008;**60**(15):1615-1626

[23] Hu C-MJ, Zhang L. Nanoparticle-based combination therapy toward overcoming drug resistance in cancer. Biochemical Pharmacology. 2012;**83**:1104-1111

[24] Yu H, Xu Z, Chen X, Xu L, Yin Q, Zhang Z, et al. Reversal of lung cancer multidrug resistance by pH-responsive micelleplexes mediating co-delivery of siRNA and paclitaxel. Macromolecular Bioscience. 2014;**14**:100-109

[25] Huang R-FS, Wei Y-J, Inbaraj BS, Chen B-H. Inhibition of colon cancer cell growth by Nano emulsions carrying gold nanoparticles and lycopene. International Journal of Nanomedicine. 2015;**10**:2823-2846

[26] Cunningham D, Atkin W, Lenz HJ, Lynch HT, Minsky B, Nordlinger B, et al. Colorectal cancer. Lancet. 2010;**375**: 1030-1047

[27] Gottlieb LS, Sternberg SS, Bond JH, Schapiro M, Panish JF, Kurtz RC, et al. Risk of colorectal Cancer in the families of patients with adenomatous polyps. The New England Journal of Medicine. 1996;**334**:82-87

[28] Thiery JP, Sleeman JP. Complex networks orchestrate epithelial–mesenchymal transitions. Nature Review in Molecular Cell Boilogy. 2006;7:131-142

[29] Mein JR, Lian F, Wang X-D. Biological activity of lycopene metabolites: Implications for cancer prevention. Nutrition Reviews. 2008;**66**:667-683

[30] Chen YJ, Inbaraj BS, Pu YS, Chen BH. Development of lycopene micelle and lycopene chylomicron and a comparison of bioavailability. Nanotechnology. 2014;**25**:155102

[31] Lu W, Zhang G, Zhang R, Flores LG, Huang Q, Gelovani JG, et al. Tumor

site-specific silencing of NF-kappaB p65 by targeted hollow gold nanosphere-mediated photothermal transfection. Cancer Research. 2010;**70**:3177-3188

[32] Leu JG, Chen SA, Chen HM, Wu WM, Hung CF, Yao YD, et al. The effects of gold nanoparticles in wound healing with antioxidant epigallocatechin gallate and alpha-lipoic acid. Nanomedicine. 2012;**8**:767-775

[33] Zhao C, Feng Q, Dou Z, Yuan W, Sui C, Zhang X, et al. Local targeted therapy of liver metastasis from Colon Cancer by Galactosylated liposome encapsulated with doxorubicin. PLoS One. 2013;**8**:e73860

[34] Neijt JP, ten Bokkel Huinink WW, van der Burg MEL, van Oosterom AT, Willemse PHB, Vermorken JB, et al. Long-term survival in ovarian cancer: Mature data from the Netherlands joint study group for ovarian cancer. European Journal of Cancer & Clinical Oncology. 1991;**27**:1367-1372

[35] Jamieson ER, Lippard SJ. Structure, recognition, and processing of cisplatin–DNA adducts. Chemical Reviews. 1999;**99**:2467-2498

[36] Ganta S, Singh A, Patel NR, Cacaccio J, Rawal YH, Davis BJ, et al. Development of EGFR targeted Nano emulsions for imaging and novel platinum therapy of ovarian cancer. Pharmaceutical Research. 2014;**31**:2490-2502

[37] Cronin M, Dearden J, Duffy J, Edwards R, Manga N, Worth A, et al. The importance of hydrophobicity and electrophilicity descriptors in mechanistically-based QSARs for toxicological endpoints. SAR QSAR Environment. 2002;**13**:167-176

[38] Ganta S, Amiji M. Coadministration of paclitaxel and curcumin in Nano emulsions formulations to overcome multidrug resistance in tumor cells. Molecular Pharmaceutics. 2009;**6**:928-939

[39] Zheng N, Gao Y, Ji H, Wu L, Qi X, Liu X, et al. Vitamin E derivative based multifunctional Nano emulsions for overcoming multidrug resistance in cancer. Journal of Drug Targeting. 2016;**24**:1-27

[40] Jemal A, Bray F, Center MM, Ferlay J, Ward E, Forman D. Global cancer statistics. CA: a Cancer Journal for Clinicians. 2011;**61**:69-90

[41] Reya T, Morrison SJ, Clarke MF, Weissman IL. Stem cells, cancer, and cancer stem cells. Nature Cell Boilogy. 2001;**414**:105-111

[42] Murali R, Varghese BA, Nair R, Konrad CV. The role of cancer stem cells in tumor heterogeneity and resistance to therapy. Canadian Journal of Physiology and Pharmacology. 2017;**95**:1-15

[43] Ahmad G, El Sadda R, Botchkina G, Ojima I, Egan J, Amiji M. Nano emulsions formulation of a novel taxoid DHA-SBT-1214 inhibits prostate cancer stem cell-induced tumor growth. Cancer Letters. 2017;**406**:71-80

[44] Gillet J-P, Calcagno AM, Varma S, Marino M, Green LJ, Vora MI, et al. Redefining the relevance of established cancer cell lines to the study of mechanisms of clinical anti-cancer drug resistance. Proceedings of the National Academy Science USA. 2011;**108**:18708-18713

[45] Botchkina GI, Zuniga ES, Das M, Wang Y, Wang H, Zhu S, et al. New-generation taxoid SB-T-1214 inhibits stem cell-related gene expression in 3D cancer spheroids induced by purified colon tumor-initiating cells. Molecular Cancer. 2010;**9**:192

[46] Jones RJ, Hawkins RE, Eatock MM, Ferry DR, Eskens FA, Wilke H, et al. A phase II open-label study of DHA-paclitaxel (Taxoprexin) by 2-h intravenous infusion in previously untreated patients with locally advanced or metastatic gastric or oesophageal adenocarcinoma. Cancer Chemotherapy and Pharmacology. 2008;**61**:435-441

[47] Moura JA, Valduga CJ, Tavares ER, Kretzer IF, Maria DA, Maranhão RC. Novel formulation of a methotrexate derivative with a lipid Nano emulsions. International Journal of Nanomedicine. 2011;**6**:2285-2295

[48] Winter E, Pizzol CD, Locatelli C, Silva AH, Conte A, Chiaradia-Delatorre LD, et al. In vitro and in vivo effects of free and Chalcones-loaded Nano emulsions: Insights and challenges in targeted Cancer chemotherapies. International Journal of Environmental Research and Public Health. 2014;**11**:10016-10035

[49] Periasamy VS, Athinarayanan J, Alshatwi AA. Anticancer activity of an ultrasonic Nano emulsions formulation of Nigella sativa L. essential oil on human breast cancer cells. Ultrasonics Sonochemistry. 2016;**31**:449-455

[50] Migotto A, Carvalho VFM, Salata GC, Da Silva FWM, Yan CYI, Ishida K, et al. Multifunctional Nano emulsions for intraductal delivery as a new platform for local treatment of breast cancer. Drug Delivery. 2018;**25**:654-667

[51] Natesan S, Sugumaran A, Ponnusamy C, Thiagarajan V, Palanichamy R, Kandasamya R. Chitosan stabilized camptothecin Nano emulsions : Development, evaluation and biodistribution in preclinical breast cancer animalmode. International Journal of Biological Macromolecules. 2017;**104**:1846-1852

[52] Kretzer IF, Maria DA, Guido MC, Contente TC, Maranhão RC, Maria D. Simvastatin increases the antineoplastic actions of paclitaxel carried in lipid Nano emulsions in melanoma-bearing mice. International Journal of Nanomedicine. 2016;**11**:885-904

[53] Favero GM, Paz JL, Otake AH, Maria DA, Caldini EG, De Medeiros RS, et al. Cell internalization of 7-ketocholesterol-containing Nano emulsions through LDL receptor reduces melanoma growth in vitro and in vivo: A preliminary report. Oncotarget. 2018;**9**:14160-14174

[54] Monge-Fuentes V, Muehlmann LA, Figueiró Longo JP, Rodrigues Silva J, Fascineli ML, de Souza P, et al. Photodynamic therapymediated by acai oil (Euterpe oleracea Martius) in Nano emulsions: A potential treatment for melanoma. Journal of Photochemistry and Photobiology B: Biology. 2017;**166**:301-310

[55] Journo-Gershfeld G, Kapp D, Shamay Y, Kopeček J, David A. Hyaluronan oligomers-HPMA copolymer conjugates for targeting paclitaxel to CD44-overexpressing ovarian carcinoma. Pharmaceutical Research. 2012;**29**:1121-1133

[56] Liebmann J, Cook J, Mitchell J. Cremophor EL, solvent for paclitaxel, and toxicity. Lancet. 1993;**342**:1428

[57] Matsubara Y, Katoh S, Taniguchi H, Oka M, Kadota J, Kohno S. Expression of CD44 variants in lung Cancer and its relationship to Hyaluronan binding. Journal of International Medicine. 2000;**28**:78-90

[58] Kim J-E, Park Y-J. Improved antitumor efficacy of hyaluronic acid-complexed paclitaxel Nano emulsions in treating non-small cell lung

Cancer. Biomolecules & Therapeutics. 2017;**25**:411-416

[59] Chang H-B, Chen B-H. Inhibition of lung cancer cells A549 and H460 by curcuminoid extracts and Nano emulsions prepared from Curcuma longa Linnaeus. International Journal of Nanomedicine. 2015;**10**:5059-5080

[60] Chen H, Zhang W, Zhu G, Xie J, Chen X. Rethinking cancer nanotheranostics. Nature Review Materials. 2017:2

[61] Fernandes DA, Fernandes DD, Li Y, Zhang Z, Rousseau D, Gradinaru CC, et al. Synthesis of stable multifunctional perfluorocarbon nano emulsions for Cancer therapy and imaging. Langmuir. 2016;**32**:10870-10880

Chapter 5

Nanoemulsions for Oil and Gas Applications

Nouf Aljabri

Abstract

As the oil and gas industry continues to evolve, the utilization of advanced materials becomes crucial for maximizing efficiency and productivity. Nanoemulsions (NEs) have emerged as a promising solution for various downhole applications. Their unique properties, enhanced stability, and improved performance have led to applications in enhanced oil recovery, drilling fluids, fracturing fluids, and produced water treatment. However, while NEs offer significant advantages, production costs, stability during transportation and storage, as well as scale-up challenges must be carefully considered. This chapter aims to provide an overview of NEs for oil and gas applications, discussing the current benchmark, potential implementation, properties, and various applications. Furthermore, it will provide recommendations and insights on how to effectively implement NEs in the field. It is important to recognize that the ongoing research and development efforts hold the potential to further revolutionize the oil and gas applications and contribute to a more sustainable processes and operations.

Keywords: EOR, asphaltene cracking, drilling fluids, anti-corrosion, scaling inhibitor, fracturing fluids, economic feasibility, scale-up

1. Introduction

Surfactants play a vital role in numerous downhole applications within the oil and gas industry. Nevertheless, they frequently face challenges related to stability when exposed to reservoir conditions. These conditions include temperature levels surpassing 70°C, pressures exceeding 3000 psi, and the presence of high ionic strength reaching up to 120,000 ppm [1, 2]. Moreover, the presence of hardness ions, such as calcium and magnesium, can form insoluble complexes with the injected chemicals (e.g., surfactants), leading to a drop in the injected chemicals' efficiency. Surfactants also exhibit the tendency to be absorbed near the well-bore owing to their interaction with rocks, which vary based on the specific surfactant and rock properties [3, 4]. One approach to overcoming this absorption is to flood the well with additional quantities, which raises concerns for both the economy and the environment. Approximately 15 years ago, literature in the oil and gas industry explored the use of nanoscale materials to enhance the performance of conventional chemicals. Since then, technologies and materials have evolved to tackle these challenges. Nanoemulsions (NEs), with their enhanced stability and unique properties, offer a potential solution to these challenges [5]. They have several important properties

IntechOpen

that contribute to their efficiency in oil and gas applications. Their small droplet size (20–200 nm) and larger interfacial area contribute to increased contact with the target substances, significantly improving performance [6]. Additionally, they exhibit unique rheological behavior and transparency, making them highly desirable in applications where precise control over fluid properties is essential [7]. A notable advantage of NEs in oil and gas applications is their significantly lower concentration requirement compared to commonly used surfactants. This characteristic plays a crucial role in reducing adsorption near the well-bore as well as the conventional Enhanced oil recovery (EOR) cost. Consequently, it brings forth numerous benefits for applications in the oil and gas sector, such as improved oil recovery, remediation of oil spills, and enhanced drilling fluids. As a result, the utilization of NEs has the potential to contribute to process optimization, enhanced environmental sustainability, and safer operational practices. Ongoing research and development in NEs technology have shown great potential for addressing future challenges in this sector. This chapter provides a review of the utilization of NEs to transform numerous oil and gas applications, showcasing the current state of the art, plausible mechanisms, and ongoing efforts to maximize NEs performance. Moreover, it discusses NEs' implementations and, more importantly, the limitations that need to be considered in order to fully unlock the potential of NEs.

2. NEs properties

Nanoemulsions have emerged as a potential game-changer in the realm of oil-field applications [8–10]. NEs, offering distinctive characteristics, effectively overcome the limitations encountered during the use of conventional surfactants and polymeric surfactant injections. This enables NEs to provide numerous advantages in various applications, as follows:

1. *Stability:* The small particle size of NEs, usually ranging between 20 and 200 nanometers, enhances stability by reducing the rate of coalescence and gravitational separation. This enhanced stability has the potential to facilitate prolonged contact between the injected fluid and the trapped hydrocarbon, ultimately leading to more effective oil recovery, cuttings decontamination, and spill remediation.

2. *Interfacial Tension:* NEs reduces the interfacial tension (IFT) between the oil and water phases. This property enables better oil displacement, as lower IFT (10-2-1 mN/m) enhances the ability of the injected fluid to spread and displace the trapped oil. Consequently, NEs show enhanced performance in reducing residual oil saturation within reservoirs.

3. *Mobility:* The nano-scaled size of NEs' droplets (10-100 nm) contributes to their low viscosity, which enables efficient flow through porous media. This increased mobility improves sweep efficiency, ensuring a higher proportion of trapped oil is recovered. NEs can effectively traverse pores and throats in geological formations, maximizing the potential for oil recovery.

4. *Compatibility:* NEs provide an eco-friendly alternative in the field of downhole applications. By minimizing the usage of surfactants and other additives, NEs reduce the environmental impact typically associated with emulsion-based methods.

5. *Improved coverage:* The nanoscale droplets offer enhanced coverage and chemical adhesion, suggesting effective prevention against corrosion and scale formation.

3. NEs applications for oil and gas applications

3.1 Enhanced oil recovery

Enhanced oil recovery (EOR) techniques have remained crucial in the oil and gas industry due to their ability to extract more hydrocarbon from swept oil fields. Customized injection processes, including thermal, chemical, and gas methods, are usually employed based on reservoir characteristics (**Figure 1**). Among these techniques, Chemical EOR (CEOR) methods, specifically surfactant flooding, have been widely used for a decade (1). However, the surfactants do have some limitations such as high viscosity, retention, and adsorption (2). To overcome these challenges, the exploration and development of NE have been pursued as they offer potential advantages over traditional surfactants for EOR. The unique properties of NEs, including reduced interfacial tension and improved viscosity control, contribute to enhancing the mobility of oil and displacing it from porous rocks. Consequently, NEs might help to significantly increase the overall recovery rate. In a recent study by Mandal and Kumar, an NE for EOR was developed using Tween-40 as an anionic surfactant-based NE [12]. This NE's formulation exhibited an average droplet size ranging from 18 to 30 nm, making it ideal for efficiently navigating and recovering trapped oil from the fine pores within oil-bearing formations [12].

In an experiment on flooding using a sand pack system, a small amount of NE slug was injected, leading to about 30% increase in recovery compared to water flooding [12]. The flooding characteristics of NEs, consisting of 0.50 wt % Tween 40 and 0.03 wt % silica nanoparticles, were also studied using a Cartesian Cuboidal Single porosity grid model. The saturation of oil decreased from around 81–86% to 21.83% and 20.70% for the NE and silica-boosted NE, respectively [12]. Modified

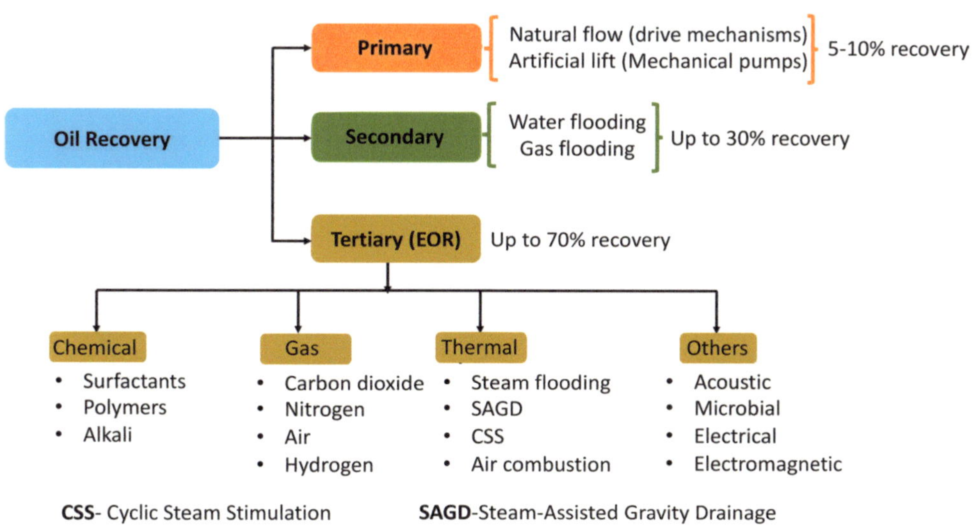

Figure 1.
Commonly used methods for EOR. Represented from Bello et al. [11].

relative permeability curves confirmed the change in wettability. Notably, the use of silica-stabilized NE resulted in a 21% reduction in oil saturation compared to a 14% reduction with conventional NE [12]. It is worth mentioning that the utilization of silica-stabilized NE has resulted in a 21% reduction in oil saturation, while the reduction is only 14% in the case of conventional NE [12].

The mechanisms behind the EOR demonstrated by NEs have been investigated and discussed extensively in literature [9]. The mechanisms through which NE enhances oil recovery include interfacial tension reduction, wettability alteration, mobility control, and emulsification. Despite the comparatively low concentrations of surfactants in the NEs formulation, these NEs function as active interfacial agents, demonstrating a reduction in IFT values. This reduction in interfacial tension helps in mobilizing trapped oil droplets and facilitates their flow toward production wells [8]. Wettability alteration is another mechanism that can enhance oil recovery by modifying the reservoir's rock wettability (**Figure 2**) [13]. NEs can adsorb onto rock surfaces, changing the wetting behavior from oil-wet to water-wet and improving oil displacement and sweep efficiency. NEs can also function as agents for controlling mobility by thickening the injected fluids, thereby improving oil recovery (IOR). Due to their properties, NEs are theorized to have the potential to mobilize highly viscous crudes and contribute to improved oil recovery. Furthermore, NEs can solubilize trapped oil within reservoir pores, increasing the overall recovery factor. However, the success of NE-based EOR depends on factors such as surfactant selection, reservoir conditions, reservoir rock properties, and injection strategies. Further research and field-scale testing are necessary to optimize the application of NE technology for enhanced oil recovery and to overcome any associated challenges.

3.1.1 Selection parameters for EOR

Selecting the appropriate surfactant variant for EOR is of utmost importance in augmenting oil extraction from reservoirs by reducing interfacial tension, modifying wettability, and enhancing fluid displacement. The use of nonionic surfactant-based NEs demonstrated an array of advantages at lab-scale, including efficient solubilization, controlled release, and improved oil transport, specifically in carbonate

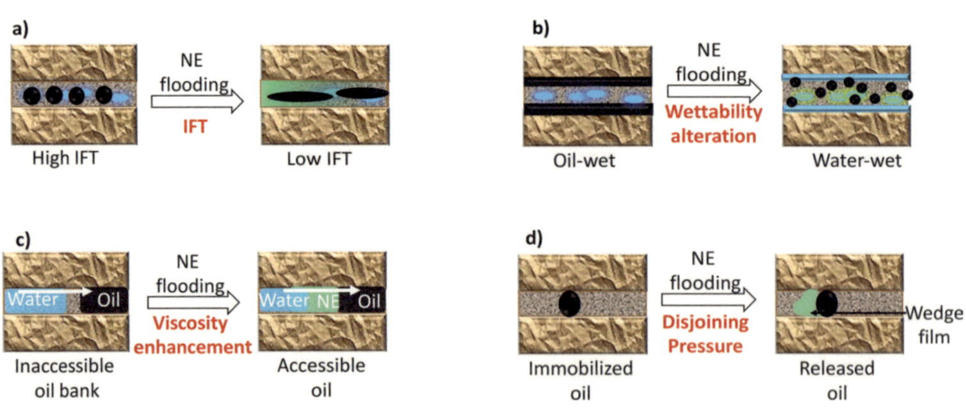

Figure 2.
A summary of the commonly proposed mechanisms for the EOR with the aid of nanoscale materials (a) IFT reduction, (b) wettability alteration, (c) viscosity enhancement, and (d) disjoining pressure. Represented with permission from [8].

formations. Results from coreflood tests have shown that NE does not agglomerate, even in rock samples with permeabilities of less than 0.1 mD [14]. This suggests that NE particles remain dispersed within the fluid and do not form clusters that could obstruct the pore spaces within the rock. Consequently, this mitigates the possibility of potential formation damage caused by agglomeration.

The hydrophilic–lipophilic balance (HLB), along with the surfactant type, is crucial in the successful design of NEs for EOR. The HLB value determines the distribution of surfactants at the oil–water interface and, as a result, determines the type of NE formed. HLB value below 6, and oil-soluble surfactants form water-in-oil (W/O) emulsions. Conversely, a value above 8 and water-soluble surfactants lead to oil-in-water (O/W) emulsions [8]. The concentration of stabilizers should also be carefully considered to ensure optimal NE stability. Recent research in the field suggests that the ideal surfactant concentration for EOR is, on average, 1.5–2 wt% [8, 10]. This facilitates the accumulation of surfactant molecules at the interface. As the thickness of the surfactant film increases, the radius of the droplets also increases, resulting in decreased flocculation.

It is of utmost importance to conduct a bottle test under reservoir conditions to verify the stability of the formulated NE for EOR. This critical test aims to assess how the produced NE reacts when exposed to formation water and reservoir temperature. To examine the performance of different formations, it is necessary to introduce either carbonate or another type of ground rock into one of the tubes, depending on the targeted lithology. Simultaneously, another tube should only include the brine and NE formulation while the third tube should be supplemented with an equivalent amount of crude oil. All three test tubes should be properly sealed and incubated in a vacuum oven set at the targeted formation temperature, preferably ≥90°C. This is to assess the performance of NE under diverse conditions. The tube containing brine and NE will reveal important information on the synergetic influence of salinity and temperature on the produced NE. Moreover, the pressure tube, which includes NE, brine, and rocks, helps us identify any potential adsorption phenomena. On the other hand, the tube containing the crude from the formation provides valuable insight into the solubilization between NE and crude. It is highly recommended that this test be conducted for at least a month with continuous monitoring to detect any visible phase separation. The monitoring process should be coupled with daily dynamic light scattering (DLS) analysis to identify any flocculation tendencies. In addition, it is essential to examine the interfacial tension (IFT) and assess the effect of the NE formulation on reducing IFT. It is preferable to aim for exceptionally low IFT values, ranging from 10^{-3} to 10^{-4} mN/m [15]. Evaluating the NE as an IFT reducer should involve conducting equilibrated IFT measurements on the incubated sample containing the crude and NE phases. To evaluate the impact of the NE on rock wettability, both static and dynamic contact angle tests should be conducted. Furthermore, addressing the correlation between droplet size, time, and temperature is crucial to control any undesirable agglomeration. Another parameter worth screening is the surface charge of the droplets, as it helps to identify the potential existence of repulsive forces between the NEs. Additionally, the in-depth analysis of fluid compatibility within the reservoir matrix becomes crucial to confirm the effectiveness of NEs in displacing trapped oil. It is recommended to employ representative core samples and fluids mimicking the targeted formation properties. When conducting a coreflood test, it is necessary to apply reservoir temperature and pressure to comprehend the behavior of the NE within the formation. Typically, brine is initially injected to establish a recovery baseline, followed by NE flooding to determine additional recovery potential. Once the coreflood is completed, the cores can be precisely sliced and

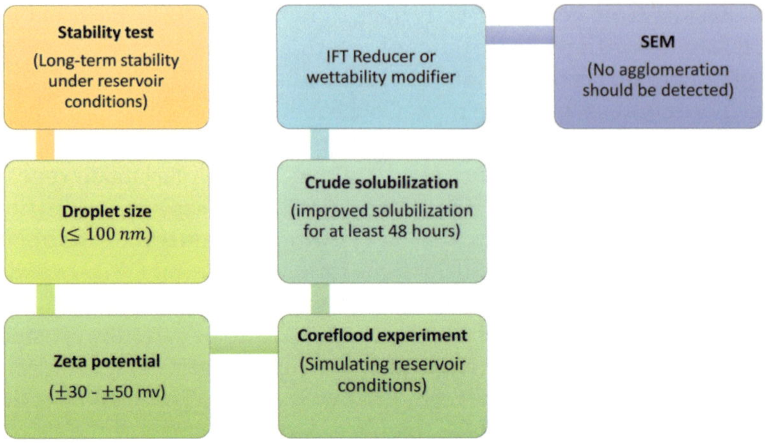

Figure 3.
Overview of different characterization techniques utilized for assessing the formulation of NEs for EOR.

characterized by a scanning electron microscopy (SEM) to identify any agglomeration that might have occurred, subsequently affecting the porosity of the cores. These studies are essential to develop a fundamental understanding of the process. **Figure 3** provides an overview of characterization techniques that can be utilized to assess the formulation of NEs when considering their application in EOR.

3.1.2 NE injection

The NE Injectivity can be significantly improved by adjusting the injection pressure. In order for the injection process to be successful and simultaneous, it is important for the bottom-hole pressure (BHP) to exceed the reservoir pressure [16]. If the BHP falls below the reservoir pressure, it creates resistance against the injected NE, which prevents effective injection [16]. However, it is essential to consider any potential negative effects of pressure on the stability of the injected NE. Therefore, conducting a coreflood experiment on cores from either the chosen formation for the trial or a similar one is recommended. This experiment should consider the influence of reservoir pressure, injection pressure, fluid compatibility, and rock petrophysics to obtain valuable insights. It is noteworthy to highlight that preserving the long-term stability of NEs in the reservoir might present a challenge. Thus, factors such as pH, divalent ions, and shear forces can lead to the breakdown of emulsions and need to be carefully addressed.

3.2 Fracturing fluids

Hydraulic fracturing is a widely utilized technique in the oil industry to enhance oil recovery from reservoirs. This method plays a vital role in extracting hydrocarbons from tight reservoirs. It involves the use of "fracturing fluids" or "frac fluids" to facilitate the creation and expansion of fractures within the reservoir rock, enabling the release of trapped hydrocarbons. NEs are explored as fracturing fluid carrier systems due to their decreased viscosity compared to conventional fluids. This characteristic allows for enhanced fluid flow, leading to reduced pumping pressures essential for fracturing operations. The improved flow features of NEs enable effective transportation and placement of proppants within fractures, thus contributing to greater well productivity. Economically point of view, NEs offer multiple advantages. Their decreased viscosity and improved fluid flow properties result in decreased requirements for slickwater (water enriched with sand and chemicals) and chemicals during fracturing operations. Consequently, the reduced volume of fluids leads to cost savings associated with water sourcing, transportation, and disposal. Furthermore, the enhanced stability and efficient proppant transportation achieved with NEs can significantly enhance well productivity, maximizing the return on investment. Zhou et al., have proposed an innovative approach to enhance hydraulic fracturing and achieve EOR simultaneously. The results demonstrated that the addition of a mere 0.10 wt% of nanoemulsion to the drag reducer (FR-900) solution resulted in a significant 12.79% increase in oil recovery. These findings emphasize the potential of utilizing this novel technique to optimize fracturing operations and improve oil extraction efficiency [17].

Moreover, the utilization of NEs can effectively mitigate the damage caused by water locking in tight sandstone reservoirs during hydraulic fracturing. Water locking refers to the adverse impact arising from the introduction of a substantial amount of water into such reservoirs. Zhou et al. conducted a study wherein they suggested a flowchart to evaluate the compatibility between fracturing fluids containing NEs and working fluids. According to the results, the incorporation of NEs into the working fluid modifies the wettability of the reservoir rocks. This alteration leads to a reduction in the absorption and retention of external fluids, thereby aiding in the alleviation of water-locking damage [18]. Nevertheless, a comprehensive evaluation is imperative due to environmental issues, including the possibility of water contamination and toxicity, prior to the widespread adoption of this solution.

3.3 Corrosion inhibitors

Corrosion refers to the deterioration of materials caused by chemical, biological, biochemical, or electrochemical reactions with the environment. It poses a challenge to oil and gas infrastructure, leading to pipeline leaks, equipment failures, and environmental contamination [19]. The financial impact of corrosion is substantial, with a high cost of repair, replacement, and maintenance annually within the oil and gas sector. It can occur through various mechanisms, such as uniform corrosion, the most common type, which involves the even degradation of a material surface. Pitting corrosion, on the other hand, leads to localized corrosion and the formation of small cavities or pits on metal surfaces. Crevice corrosion usually takes place in the gaps between tubulars, while galvanic corrosion arises when dissimilar metals come into contact [20]. A relatively recent concern in the oil and gas industry is the concept of terbo-corrosion. Terbo-corrosion is a term that describes the interaction of corrosion with turbulent flow conditions. It occurs primarily in pipelines, where the high velocity of oil or gas facilitates the erosion of protective barriers and promotes corrosion [21]. It can result in rapid deterioration and heightened vulnerability of infrastructure, demanding focused attention to corrosion control strategies. Thus, the corrosion control methods are applied to overcome the corrosion adverse impact.

Anti-corrosion fluids play a crucial role in preventing and mitigating corrosion. One commonly used example is corrosion inhibitors, which act by forming a protective layer on the metal surface, thereby inhibiting the electrochemical reactions that cause corrosion. Another example is oxygen scavengers, which reduce the amount of dissolved oxygen in the fluid, minimizing the possibility of corrosion [22]. Furthermore, biocides are used to control the growth of microorganisms that can contribute to microbial corrosion. These anti-corrosion fluids control the corrosion via one of the following mechanisms: (1) barrier protection, (2) inhibitor, and (3) passivation [23, 24]. The barrier protection to create a barrier between the pipes or tubulars and the environmental trigger. This barrier prevents the contact of corrosive agents with the metal, thereby minimizing the reactions that lead to corrosion [25]. The corrosion inhibitors are usually used as additives to the drilling fluids to impede the corrosion process. These inhibitors can neutralize or reduce the activity of corrosive species, slowing down the rate of corrosion. They achieve this by forming a protective layer on the metal surface or by altering the electrochemical reactions involved in the corrosion process. Lastly, the passivation involves the formation of a thin layer of oxide on the metal surface, which acts as a protective film. This oxide layer is usually formed by the reaction of the metal with certain components in the fluid, creating a stable and corrosion-resistant surface [26, 27].

While anti-corrosion fluids provide significant benefits in combating corrosion, they have certain limitations that need to be considered. For instance, the long-term effectiveness of corrosion inhibitors may diminish over time. Additionally, the inhibitors' performance can vary depending on the specific environment or fluid composition. Oxygen scavengers may also require careful monitoring as they can deplete quickly, necessitating regular replenishment. Moreover, some anti-corrosion fluids may have adverse environmental impacts, requiring proper handling and disposal [26, 28]. Therefore, the anti-corrosion encapsulation in NE was suggested to prolong its anti-corrosion efficiency. The nanosized droplets enable a greater surface coverage, resulting in improved protection against corrosive agents. Moreover, the incorporation of corrosion inhibitors actively impedes the corrosion process, extending the lifespan of metallic structures and reducing maintenance costs. The efficacy of Shirazi thyme, a green corrosion inhibitor, in acidic media, was investigated by encapsulating it in NE droplets measuring 20 nm in size [29]. The encapsulated thyme extract demonstrated a corrosion inhibition efficiency of 94%, surpassing the 76% efficiency of using thyme without encapsulation. Surprisingly, a mere concentration of 40 ppm proved sufficient to achieve an impressive 94% efficiency, in contrast to the requirement of 800 ppm without the process of encapsulation. The unencapsulated thyme extract adheres to the surface through physical adsorption, while the encapsulated thyme tends to engage in a physical-chemical interaction with the surface. Interestingly, the unencapsulated thyme extract displayed adsorption of free energy equal to $-23\ kJ{\cdot}mol^{-1}$, whereas the encapsulated extract in NEs exhibited improved adsorption of $-30\ kJ{\cdot}mol^{-1}$ [29]. The hypothesis suggests that the sodium dodecyl sulfate (anionic surfactant, SDS) present in the formulation exhibits a strong inclination to be adsorbed onto the metals. Once adsorbed, the micelles break apart to release the thyme. However, more research is deemed necessary to tackle obstacles associated with cost-effectiveness, assessment of environmental impact, and seamless incorporation into existing industry practices.

3.4 Anti-scaling

Scales form due to the precipitation of minerals found in fluids present in oil and gas reservoirs [30]. These minerals, such as calcium carbonate, barium sulfate, and iron sulfide, can attach themselves to various production equipment, including pipelines, well-bore completions, and surface facilities [31–33]. This scale deposition not only influences flow efficiency but also disrupts flow dynamics, potentially resulting in reduced production rates, increased energy consumption, and, ultimately, diminished economic gains [34]. In addition to negatively impacting production, scaling presents significant operational challenges. Scale formations have the potential to obstruct or completely block pipelines, leading to costly production shutdowns that may require cleaning or replacement of affected tubulars. As a result, it becomes essential to employ anti-scaling measures to ensure the preservation and effectiveness of oil and gas production systems. Earlier methods such as chemical inhibitors, dissolvers, and mechanical treatments have demonstrated partial effectiveness in combating scaling [33, 35, 36]. However, these methods often encounter disadvantages such as exorbitant expenses, limited efficiency, and environmental apprehensions. Hence, there arises a necessity for developing new methodologies and materials that offer remarkable performance compared to conventional approaches. There are multiple advantages associated with NEs-based anti-scalants and corrosion inhibitors compared to traditional methods. First, their small droplet size enables better dispersion and coverage of the active ingredients, thereby enhancing their overall effectiveness. Second, the NEs possess a controlled release mechanism that potentially ensures a sustained and prolonged action, thereby minimizing the need for frequent treatment. This characteristic not only improves the treatment efficiency but also reduces the operational costs involved in monitoring and maintenance tasks. Additionally, the composition of NEs can be tailored to target specific scale-forming minerals, allowing for a more precise and targeted approach. By customizing the formulation, the inhibitory effect can be optimized for diverse applications, resulting in higher inhibition efficiency and reduced chemical consumption. Furthermore, the low dosage requirements of NEs-based anti-scalants make them economically viable and environmentally friendly alternatives. Del et al. presented an innovative delivery system known as phosphino-poly (carboxylic acid) (PPCA) based NEs. This delivery vehicle exhibits significant potential in efficiently handling mineral scale accumulation in oil reservoirs, particularly those that are depleted or prone to water sensitivity [37]. The stability of NEs with a PPCA content of up to 3% was found to be satisfactory over an 8-hour span at 100°C [37]. To assess the injectivity and absorption of PPCA NE, a sandpack test was conducted on porous media. The results indicated that only 4% of the total formulation was adsorbed, confirming the anti-scale inhibitor encapsulation efficiency [37]. A separate study conducted by Luo et al. demonstrated the production of degradable anti-scaling NEs using orthogonal experiments as an experimental approach [38]. The results obtained indicated an inhibition rate of over 89.6%, exhibiting no precipitation with the formation water. Mady and Kelland have extensively reviewed the influence of nanotechnology including the NEs on formation control and other downhole applications [39]. With the continuous progress of research in this domain, we anticipate additional breakthroughs in NE technology, which will pave the way for enhanced and multifaceted approaches to corrosion control and anti-scaling.

3.5 Asphaltene cracking

Asphaltene, a heavy molecule present in crude oil, is a diverse blend of aromatic hydrocarbons [40]. Its intricate structure and considerable molecular weight hinder its solubility in traditional hydrocarbon solvents. Consequently, when mixing crude oil with lighter hydrocarbons or gas during the oil production process, the precipitation of asphaltene can occur, which poses significant challenges for extraction [41]. This deposition of asphaltene leads to reduced production rates, pipeline obstructions, and equipment damage, thereby creating substantial hurdles for the oil and gas industry. These intricate organic compounds tend to aggregate and can precipitate and attach themselves in specific conditions, resulting in solid formations that impede oil flow, diminish reservoir permeability, and cause operational inefficiencies [41]. Thermal, mechanical, and chemical methods are employed as standard approaches to alleviate asphaltene deposition [42]. Thermal cracking is a process that involves subjecting asphaltenes to elevated temperatures, resulting in the breakdown of complex structures into smaller, more manageable fragments [43]. Mechanical techniques, on the other hand, rely on physical forces such as agitation, shear stress, or ultrasound to disrupt the aggregated asphaltenes [44]. Chemical cracking methods employ various solvents or additives to dissolve or disperse the asphaltenes [45]. When compared to conventional methods, NEs offer several advantages for asphaltene cracking. The small droplet size and large interfacial area of NEs greatly enhance the contact between asphaltene molecules and cracking agents, leading to faster and more efficient cracking reactions. Additionally, the dispersion of cracking agents within the emulsion results in a more uniform distribution, thus increasing the effectiveness of asphaltene cracking throughout the targeted zone.

Several case studies and research findings have provided substantial evidence supporting the potential of NEs in the cracking of asphaltenes. An example includes a study conducted by Alhreez and Wen, where they successfully demonstrated the controlled release of the asphaltene inhibitor from NE as an effective cracking agent in a laboratory-scale asphaltene deposition system [46]. The outcome of this study showcased a significant reduction in deposition tendencies. Additionally, Alhreez and Wen have recently pioneered a new approach that employs NEs as carriers for asphaltene inhibitors. The outcomes of their experiments, conducted using X-ray Diffraction (XRD) and Transmission Electron Microscopy (TEM), further supported the theory that the release of dodecylbenzene sulfonic acid (DBSA) from the NE destabilizes asphaltenes [47]. The process led to an augmentation in the separation between aromatic rings and aliphatic chains, accompanied by a decrease in the size of aromatic sheets and clusters. These alterations in the structure caused a decline in the aromaticity of asphaltene, resulting in enhanced solubility of aromatic compounds and improved release of asphaltene aggregates. **Figure 4** illustrates the likely mechanism of NE adsorption on the asphaltene sheets, followed by the subsequent release of DBSA and its impact on asphaltene [47]. The efficiency of asphaltene cracking controlled release was evaluated by Alhareez et al. where the findings demonstrate that the controlled release system effectively reduced asphaltene content, achieving an inhibition rate of approximately 84%, while also exhibiting an elongated release profile [48]. The data suggest that the destabilization of asphaltene particles is primarily caused by the release kinetics of dodecylbenzenesulfonic acid (DBSA) from NEs. This mechanism entails several steps, beginning with the adsorption of asphaltene aggregates or clusters at the water-in-oil (w/o) interface. Subsequently, DBSA NEs adsorb onto the asphaltenes both within the bulk phase and at the w/o interface. Finally, DBSA gradually releases from the NEs, resulting in the fragmentation of larger asphaltene clusters into smaller nanosized particles. As a result, this process yields a well-dispersed solution [48]. To advance in this field, it is essential to dedicate research efforts toward formulating strategies that facilitate the expansion and execution of operations in oil and gas fields.

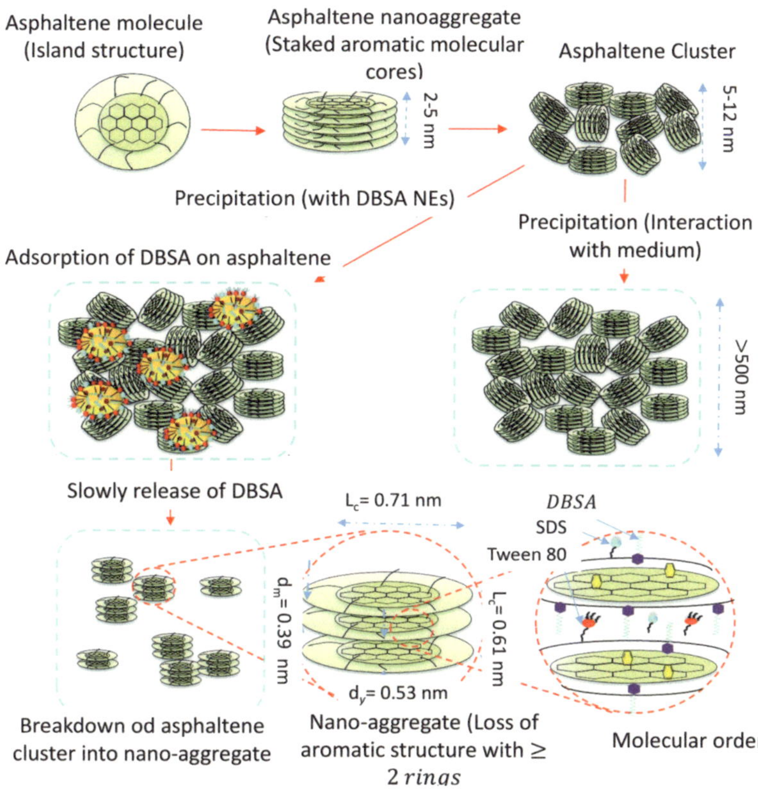

Figure 4.
Illustration demonstrates the slow release of the asphaltene inhibitor from nanoemulsion to destabilize the asphaltene. Represented from Alhareez and Wen [47].

3.6 Drill cuttings decontamination

The drilling process heavily relies on drilling fluids or mud to serve various purposes, including the cooling of the drill bit, transportation of cuttings to the surface, and management of well pressure. Nonetheless, when these drilling fluids come into contact with geological formations, they tend to become contaminated with solid particles, hydrocarbons, heavy metals, and other hazardous substances [49]. Consequently, the generated drilling cuttings, which consist of rock fragments and solids, also become contaminated. To mitigate the environmental impact associated with drilling activities, it becomes crucial to decontaminate the drilling cuttings. The main objective of this process revolves around removing contaminants from the cuttings while also recuperating any reusable or recyclable materials. By adopting this approach, the potential harm inflicted on the surrounding ecosystems gets minimized, ensuring compliance with environmental regulations. Thermal desorption stands out as a commonly employed technique for drilling-cutting decontamination [50]. Thermal desorption provides numerous benefits in terms of efficiency and effectiveness. It has the capability to process large quantities of cuttings, and the resulting soil or rock is often suitable for reuse or proper disposal in accordance with environmental regulations. Nevertheless, it is crucial to meticulously manage the energy consumption and emissions associated with this procedure to minimize its carbon footprint and overall environmental impact, as well as ensure scalability. Besides thermal desorption, other techniques are available for the decontamination of drilling cuttings, such as solvent extraction and solidification/stabilization [51]. Solvent extraction involves the use of solvents to dissolve and separate contaminants from the cuttings, while solidification/stabilization entails the mixing of binders and additives with the cuttings to immobilize the contaminants and prevent their migration extraction [52]. The incorporation of NEs in the decontamination process of drilling-cutting fluids offers numerous advantages, which makes it a highly promising solution for preserving the cleanliness and effectiveness of these fluids [53]. NEs exhibit exceptional dispersing properties that facilitate the dissolution and dispersion of both organic and inorganic pollutants typically found in drilling-cutting fluids. By leveraging the minute droplet size of NEs, organic contaminants like hydrocarbons can be efficiently disintegrated, thereby enhancing their solubility [53]. In addition, NEs prove effective in dispersing inorganic pollutants, such as heavy metals and salts, effectively preventing their accumulation and mitigating potential harm to drilling equipment. The utilization of NE has been successfully demonstrated in decontaminating drilling cuttings, as evidenced by the results indicating its ability to modify wettability and facilitate the desorption of oil and other contaminants from rock surfaces [54].

Lim et al. investigated the potential application of NEs for the removal of oil-based mud and mud cake during well-bore cleaning [55]. The experiments encompassed a range of factors such as temperature, pressure, electrolytes, bridging agents, acidity, and polymers. The cleaning process involved several steps, including the reduction of interfacial tension, emulsification and solubilization of oil, improvement in wettability, and efficient separation of oil from water and solids. The findings of the study clearly indicated the high effectiveness of NEs in the cleaning of mud and cuttings, regardless of the tested conditions. Interestingly, results were obtained within a time frame of less than 10 minutes [55]. This discovery emphasizes the immense possibilities of NEs within the oil and gas sector, particularly concerning sustainable methods and operational efficacy. Given the industry's persistent drive toward environmentally friendly and efficient practices, the adoption of NEs for drill-cutting decontamination stands out as a promising and vital advancement. As the industry perseveres in its pursuit of sustainability and operational efficiency, the integration of NEs in drill-cutting decontamination proves to be a promising and essential innovation.

3.7 Well-bore stabilization

Shale hydration pertains to the process by which shale formations in the oil and gas industry absorb water [56]. When shale comes into contact with water-based drilling fluids, it tends to absorb water, resulting in swelling and weakening of the rock structure. Consequently, this can lead to damage in the formation, diminished stability of the well-bore, and ultimately hinder drilling and production activities [56]. Numerous factors such as pressure, temperature, and mineral composition play a role in shale hydration. The effective management of shale hydration holds great importance in ensuring successful drilling and completion operations in shale formations. As a potential solution to tackle well-bore instability problems caused by shale hydration, researchers are exploring the use of NEs containing soluble silicate and organosilanes. These NEs establish a hydrophobic barrier on shale surfaces, acting as a protective shield during high-temperature drilling. Furthermore, in shale dispersion tests, the NEs exhibited a remarkable shale recovery rate of 106.4%, surpassing that achieved by water at a rate of 20% [57]. This study indicates that NEs have the ability to ensure the stability of well-bores in shale formations with great efficiency. Additionally, the contact angles of diiodomethane on shale samples were observed to be below 2° and slightly increased as the shale cuttings acquired more hydrophobic properties [57]. This observation signifies that diiodomethane droplets swiftly expanded upon contact with the samples, highlighting a significant affinity toward oil. The potential of NEs in enhancing drilling effectiveness and averting well-bore impairment has been established.

Utilizing NEs at high temperatures (≥ 150°C) can yield a compact, superhydrophobic coating on ceramic sheets, resulting in notable advantages. This coating effectively minimizes filtration loss by 84.3% and enhances the robustness of shale by 2.86% following immersion in the NE at 180°C for 48 hours [58]. Furthermore, it significantly diminishes water absorption in shale cuttings, thereby elevating water resistance by 91.56%. The obtained data demonstrates thermal stability and dynamic water repellency even at elevated temperatures. The composition of this film comprises modified silica structures, namely nanofilaments and nanospheres [58]. The ongoing research in the field is dedicated to exploring advanced imaging techniques, surface modification strategies, and improved modeling approaches in order to gain a comprehensive understanding and optimize the wettability of well-bores [58].

4. Limitations and areas of improvements

The discovery highlights the significant potential of NEs in the oil and gas industry, particularly in terms of sustainable practices and operational efficiency. While promising findings are often observed in laboratory-scale experiments, implementing NEs for field use poses additional challenges. One such challenge is the formulation of NEs at the rig site, where limited infrastructure for chemical production and NE preparations creates obstacles. However, the primary hurdle lies in maintaining the stability and performance of the NEs over long distances and extended periods of time. Proper attention must be devoted to the transportation, storage, and injection of NEs into reservoirs to prevent droplet aggregation and phase separation. Hence, the development of robust techniques for inline characterization and injection becomes crucial to optimize particle size and parameters like flow rate and pressure, thereby augmenting the success of scaled-up applications. The development of robust and long-lasting NEs capable of withstanding extreme temperatures, pressures, and salinities presents an additional challenge. Furthermore, the selection of appropriate constituents for NE formulations is crucial to ensure their stability at minimized costs. In addition, conducting extensive field trials and testing is highly recommended to gather sufficient data on NE performance in downhole environment. Recent research efforts have prioritized enhancing the stability, scalability, and environmental aspects of NE technology. The following recommendations can be considered for the successful field implementation.

- *Comprehensive Feasibility Studies:* Conduct thorough feasibility studies to evaluate the applicability and potential benefits of NE technology in specific reservoirs and production processes.

- *Testing and Validation:* Prioritize pilot-scale testing and validation before full-scale implementation. This ensures a proper understanding of the technology's behavior and helps identify any potential setbacks early on.

- *Regular Monitoring and Optimization:* Continuously monitor and optimize the performance of NE applications to ensure their efficiency and effectiveness in different operational conditions.

5. Conclusion

In summary, NEs application in the oil and gas production process holds immense promise for improving operational efficiency and performance. This advanced formulation, created by finely dispersing oil and water using surfactants, stabilizers, and high or low-energy mixing techniques, allows for enhanced oil displacement and increased sweep efficiency, leading to a potential improvement in production rates. Moreover, the small droplet size of the NEs provides a larger interfacial area, enabling better contact with the reservoir rock and creating an effective displacement of oil. The use of NEs also offers environmental benefits, as it reduces the needed chemical quantitates compared to the utilization of pristine chemicals. However, one of the key setbacks in utilizing NEs is maintaining their stability and compatibility with other chemicals commonly used in oil-field operations. Nanoemulsion formulations must ensure that they remain stable under harsh conditions, such as high temperatures and salinity, to ensure their effectiveness. Additionally, compatibility with existing chemicals and hydrocarbons in the oil and gas reservoirs is crucial to prevent any negative interactions that may hinder the desired outcomes. Further research and development are necessary to optimize the formulation and application of NEs in different reservoir conditions. By addressing setbacks and implementing recommended strategies, the industry can unlock the full potential of NEs, leading to improved progress in the field of oil and gas production.

Conflict of interest

The authors declare no conflict of interest.

Author details

Nouf Aljabri
KAUST Upstream Research Center, EXPEC ARC, Saudi Aramco, Thuwal, Saudi Arabia

*Address all correspondence to: noufm.aljabri@gmail.com

References

[1] Schramm L. Surfactants: Fundamentals and Applications in the Petroleum Industry. Vol. 16. United Kingdom: PALAIOS; 2010

[2] Sharifi A, Miri R, Riazi M. A holistic review of harsh conditions resistant surfactants for enhanced oil recovery in dense carbonate reservoir. Fuel. 2023;**353**:129109. Available from: https://www.sciencedirect.com/science/article/pii/S0016236123017234

[3] Ko KM, Chon BH, Jang SB, Jang HY. Surfactant flooding characteristics of dodecyl alkyl sulfate for enhanced oil recovery. Journal of Industrial and Engineering Chemistry. 2014;**20**(1):228-233. Available from: https://www.sciencedirect.com/science/article/pii/S1226086X13001494

[4] Kazemzadeh Y, Ismail I, Rezvani H, Sharifi M, Riazi M. Experimental investigation of stability of water in oil emulsions at reservoir conditions: Effect of ion type, ion concentration, and system pressure. Fuel. 2019;**243**:15-27. Available from: https://www.sciencedirect.com/science/article/pii/S0016236119300742

[5] Saxena N, Kumar A, Mandal A. Applications of nanoemulsions in EOR. In: Saxena N, Kumar A, Mandal A, editors. Nano Emulsions in Enhanced Oil Recovery. Cham: Springer International Publishing; 2022. pp. 43-46. DOI: 10.1007/978-3-031-06689-4_6

[6] Rigano L, Lionetti N. Chapter 6—Nanobiomaterials in galenic formulations and cosmetics. In: Grumezescu AM, editor. Nanobiomaterials in Galenic Formulations and Cosmetics [Internet]. USA: Elsevier; 2016. pp. 121-148. Available from: https://www.sciencedirect.com/science/article/pii/B9780323428682000061

[7] Kumar N, Mandal A. Surfactant stabilized oil-in-water nanoemulsion: Stability, interfacial tension, and rheology study for enhanced oil recovery application. Energy & Fuels. 2018;**32**(6):6452-6466. DOI: 10.1021/acs.energyfuels.8b00043

[8] Aljabri NM, Shi N, Cavazos A. Nanoemulsion: An emerging technology for oilfield applications between limitations and potentials. Journal of Petroleum Science and Engineering. 2022;**208**:109306. Available from: https://www.sciencedirect.com/science/article/pii/S0920410521009591

[9] Hou X, Sheng JJ. Properties, preparation, stability of nanoemulsions, their improving oil recovery mechanisms, and challenges for oil field applications—A critical review. Geoenergy Science and Engineering. 2023;**221**:211360. Available from: https://www.sciencedirect.com/science/article/pii/S2949891022000483

[10] Kumar N, Verma A, Mandal A. Formation, characteristics and oil industry applications of nanoemulsions: A review. Journal of Petroleum Science and Engineering. 2021;**206**:109042. Available from: https://www.sciencedirect.com/science/article/pii/S0920410521006999

[11] Bello A, Ivanova A, Cheremisin A. Foam EOR as an optimization technique for gas EOR: A comprehensive review of laboratory and field implementations. Energies. 2023;**16**:972

[12] Kumar N, Pal N, Mandal A. Nanoemulsion flooding for enhanced

oil recovery: Theoretical concepts, numerical simulation and history match. Journal of Petroleum Science and Engineering. 2021;**202**:108579. Available from: https://www.sciencedirect.com/science/article/pii/S0920410521002394

[13] Rezaei A, Abdollahi H, Derikvand Z, Hemmati-Sarapardeh A, Mosavi A, Nabipour N. Insights into the effects of pore size distribution on the flowing behavior of carbonate rocks: Linking a nano-based enhanced oil recovery method to rock typing. Nanomaterials. 2020;**10**(5):972

[14] Liang T, Li Q, Liang X, Yao E, Wang Y, Li Y, et al. Evaluation of liquid nanofluid as fracturing fluid additive on enhanced oil recovery from low-permeability reservoirs. Journal of Petroleum Science and Engineering. 2018;**168**:390-399. Available from: https://www.sciencedirect.com/science/article/pii/S092041051830367X

[15] Kumar N, Mandal A. Wettability alteration of sandstone rock by surfactant stabilized nanoemulsion for enhanced oil recovery—A mechanistic study. Colloids and Surfaces A: Physicochemical and Engineering Aspects. 2020;**601**:125043. Available from: https://www.sciencedirect.com/science/article/pii/S0927775720306361

[16] Odi U. Incorporation of homogenizer in nanoemulsion injection scheme for enhanced oil recovery. Improved Oil and Gas Recovery. 2018;**2**:1-18

[17] Wang J, Guo P, Jiang H, Zhou F. A novel multifunction fracturing fluid compounding of nano-emulsion and viscous slickwater for unconventional gas and oil. Arabian Journal of Chemistry. 2022;**15**(5):103749. Available from: https://www.sciencedirect.com/science/article/pii/S187853522200065X

[18] Wang J, Li Y, Zhou F, Yao E, Zhang L, Yang H. Study on the mechanism of Nanoemulsion removal of water locking damage and compatibility of working fluids in tight sandstone reservoirs. ACS. Omega. 2020;**5**(6):2910-2919. DOI: 10.1021/acsomega.9b03744

[19] Ma IAW, Ammar S, Kumar SSA, Ramesh K, Ramesh S. A concise review on corrosion inhibitors: Types, mechanisms and electrochemical evaluation studies. Journal of Coatings Technology and Research. 2022;**19**(1):241-268. DOI: 10.1007/s11998-021-00547-0

[20] Marya M. A Comparison of the Pitting and Crevice Corrosion of Directional Drilling Alloys with Oilfield Production Alloys. USA: Association for Materials Protection and Performance (AMMP); 2020. p. NACE-2020-14432

[21] Palimi MJ, Tang Y, Wu M, Alvarez V, Ghavidel M, Kuru E, et al. Improve the tribo-corrosion behavior of oil-in-water emulsion-based drilling fluids by new derivatives of fatty acid-based green inhibitors. Tribology International. 2022;**174**:107723. Available from: https://www.sciencedirect.com/science/article/pii/S0301679X22002961

[22] Al Helal A, Ajjaoui S, Mohammed MM, Barificani A, Znad H. Enhancing oxygen scavenging properties of erythorbic acid through transition metal catalysis. Process Safety and Environmental Protection. 2023;**175**:516-523. Available from: https://www.sciencedirect.com/science/article/pii/S0957582023004494

[23] Askari M, Aliofkhazraei M, Jafari R, Hamghalam P, Hajizadeh A. Downhole corrosion inhibitors for oil and gas production—A review. Applied Surface Science Advances. 2021;**6**:100128. Available from: https://www.

sciencedirect.com/science/article/pii/S266652392100074X

[24] Richter S, Achour M, Addis K, Singer M, Nesic S. Development and Application of a Downhole Corrosion Prediction Model. USA: Association for Materials Protection and Performance (AMMP); 2016. p. NACE-2016-7827

[25] Dong R, Yu Q, Bai Y, Wu Y, Ma Z, Zhang J, et al. Towards superior lubricity and anticorrosion performances of proton-type ionic liquids additives for water-based lubricating fluids. Chemical Engineering Journal. 2020;**383**:123201. Available from: https://www.sciencedirect.com/science/article/pii/S1385894719326130

[26] Lehmann M, Lepkova K, Pojtanabuntoeng T, Ghodkay V, Greenwood A, Hirano S, et al. Use of Oxygen Scavenger in Well Safety Valve Balancing Operations. USA: Offshore Technology Conference; 2022. p. D041S038R006. DOI: 10.4043/31509-MS

[27] Al-Janabi YT. An overview of corrosion in oil and gas industry. In: Saji S, Umoren SA, editors. Corrosion Inhibitors in the Oil and Gas Industry. USA: John Wiley and Sons; 2020. pp. 1-39. DOI: 10.1002/9783527822140.ch1

[28] Mubarak G, Verma C, Barsoum I, Alfantazi A, Rhee KY. Internal corrosion in oil and gas wells during casings and tubing: Challenges and opportunities of corrosion inhibitors. Journal of the Taiwan Institute of Chemical Engineers. 2023;**150**:105027. Available from: https://www.sciencedirect.com/science/article/pii/S1876107023003565

[29] Hassannejad H, Barati A, Nouri A. The use of nanoemulsion-based strategies to improve corrosion inhibition efficiency of thyme-based inhibitor. Journal of Molecular Liquids. 2019;**296**:112110. Available from: https://www.sciencedirect.com/science/article/pii/S0167732219307834

[30] Zhang Y, Shaw H, Farquhar R, Dawe R. The kinetics of carbonate scaling—Application for the prediction of downhole carbonate scaling. Journal of Petroleum Science and Engineering. 2001;**29**(2):85-95. Available from: https://www.sciencedirect.com/science/article/pii/S0920410500000954

[31] Vetter OJG, Phillips RC. Prediction of deposition of calcium Sulfate scale under Down-hole conditions. Journal of Petroleum Technology. 1970;**22**(10):1299-1308. DOI: 10.2118/2620-PA

[32] BinMerdhah AB, Yassin AAM, Muherei MA. Laboratory and prediction of barium sulfate scaling at high-barium formation water. Journal of Petroleum Science and Engineering. 2010;**70**(1):79-88. Available from: https://www.sciencedirect.com/science/article/pii/S0920410509002198

[33] Chen T, Wang Q, Chang FF, Al-Janabi YT. New Developments in Iron Sulfide Scale Dissolvers. USA: Association for Materials Protection and Performance (AMMP); 2016. p. NACE-2016-7264

[34] Muryanto S, Bayuseno AP, Ma'mun H, Usamah M, Jotho. Calcium carbonate scale formation in pipes: Effect of flow rates, temperature, and malic acid as additives on the mass and morphology of the scale. Procedia Chemistry. 2014;**9**:69-76. Available from: https://www.sciencedirect.com/science/article/pii/S1876619614000102

[35] Khormali A, Petrakov DG, Afshari Moein MJ. Experimental analysis of calcium carbonate scale formation and

inhibition in waterflooding of carbonate reservoirs. Journal of Petroleum Science and Engineering. 2016;**147**:843-850. Available from: https://www.sciencedirect.com/science/article/pii/S0920410516305101

[36] Olajire AA. A review of oilfield scale management technology for oil and gas production. Journal of Petroleum Science and Engineering. 2015;**135**:723-737. Available from: https://www.sciencedirect.com/science/article/pii/S0920410515301066

[37] Del Gaudio L, Bortolo R, Lockhart TP. Nanoemulsions: A New Vehicle for Chemical Additive Delivery. USA: Society of Petroleum Engineering; 2007. p. SPE-106016-MS. DOI: 10.2118/106016-MS

[38] Luo M, Sun H, Jia Z, Wen Q, Liao L. Preparation and Performance of Environment-Friendly Nanoemulsion with Antiscaling. USA: Society of Petroleum Engineers; 2012. p. SPE-152688-MS. DOI: 10.2118/152688-MS

[39] Mady MF, Kelland MA. Review of nanotechnology impacts on oilfield scale management. ACS Applied Nano Materials. 2020;**3**(8):7343-7364. DOI: 10.1021/acsanm.0c01391

[40] Ramirez-Corredores MM. Chapter 2 - Asphaltenes. In: Ramirez-Corredores MM, editor. The Science and Technology of Unconventional Oils. Amsterdam: Academic Press; 2017. pp. 41-222. Available from: https://www.sciencedirect.com/science/article/pii/B9780128012253000024

[41] Abdelazim A, Abu El Ela M, El-Banbi A, Sayyouh H. Successful approach to mitigate the asphaltenes precipitation problems in ESP oil wells. Journal of petroleum exploration and production. Technology. 2022;**12**(3):725-741. DOI: 10.1007/s13202-021-01335-7

[42] Gharbi K, Benyounes K, Khodja M. Removal and prevention of asphaltene deposition during oil production: A literature review. Journal of Petroleum Science and Engineering. 2017;**158**:351-360. Available from: https://www.sciencedirect.com/science/article/pii/S0920410517306885

[43] de Andrade Brüning IMR, Martinez ÂV, Travalloni AML. Identification of Asphaltene Deposition in Camorim Field, Northeastern Brazil. USA: Society of Petroleum Engineering; 1990. p. SPE-21102-MS. DOI: 10.2118/21102-MS

[44] Akbar SH, Saleh AA. A Comprehensive Approach to Solve Asphaltene Deposition Problem in some Deep Wells. USA: Society of Petroleum Engineering; 1989. p. SPE-17965-MS. DOI: 10.2118/17965-MS

[45] Ghloum EF, Rashed AM, Safa MA, Sablit RC, Al-Jouhar SM. Mitigation of asphaltenes precipitation phenomenon via chemical inhibitors. Journal of Petroleum Science and Engineering. 2019;**175**:495-507. Available from: https://www.sciencedirect.com/science/article/pii/S092041051831180X

[46] Alhreez M, Wen D. Controlled releases of asphaltene inhibitors by nanoemulsions. Fuel. 2018;**234**:538-548. Available from: https://www.sciencedirect.com/science/article/pii/S0016236118311293

[47] Alhreez M, Wen D. Molecular structure characterization of asphaltene in the presence of inhibitors with nanoemulsions. RSC Advances. 2019;**9**(34):19560-19570. DOI: 10.1039/C9RA02664A

[48] Alhreez M, Xiao X, Wen D. Kinetic study of controlled Asphaltene inhibitor release from Nanoemulsions. Langmuir. 2019;**35**(33):10795-10807. DOI: 10.1021/acs.langmuir.9b00481

[49] Ball AS, Stewart RJ, Schliephake K. A review of the current options for the treatment and safe disposal of drill cuttings. Waste Management & Research. 2011;**30**(5):457-473. DOI: 10.1177/0734242X11419892

[50] Okeke P, Obi C. Treatment of oil drill cuttings using thermal desorption technique. ARPN Journal of Systems and Software. 2013;**3**(7):153-158

[51] Poyai T, Getwech C, Dhanasin P, Punyapalakul P, Painmanakul P, Chawaloesphonsiya N. Solvent-based washing as a treatment alternative for onshore petroleum drill cuttings in Thailand. Science of the Total Environment. 2020;**718**:137384. Available from: https://www.sciencedirect.com/science/article/pii/S0048969720308949

[52] Hu Y, Chen X, Mu S, Li Q. Extraction and separation of petroleum pollutants from oil-based drilling cuttings using methanol/n-hexane solvent. Process Safety and Environmental Protection. 2022;**168**:760-767. Available from: https://www.sciencedirect.com/science/article/pii/S0957582022009144

[53] Denney D. Nanoemulsion-enhanced treatment of oil-contaminated cuttings. Journal of Petroleum Technology. 2013;**65**(11):125-127. DOI: 10.2118/1113-0125-JPT

[54] Saphanuchart W, Loke YS, See CH. Nanoemulsion-enhanced treatment of oil-contaminated oil-based drill solids. Journal of Petroleum Technology (USA: Society of Petroleum Engineers). 2012:SPE-162401-MS. DOI: 10.2118/162401-MS

[55] See CH, Saphanuchart W, Nadarajan S, Lim CN. NanoEmulsion for Non-Aqueous Mud Removal in Wellbore. USA: Society of Petroleum Engineers; 2011. p. SPE-149088-MS. DOI: 10.2118/149088-MS

[56] Barati P, Shahbazi K, Kamari M, Aghajafari A. Shale hydration inhibition characteristics and mechanism of a new amine-based additive in water-based drilling fluids. Petroleum. 2017;**3**(4):476-482. Available from: https://www.sciencedirect.com/science/article/pii/S2405656116301791

[57] Li Y, Wang M, An Y, Li K, Wei Z, Bo K, et al. In situ shale wettability regulation using sophisticated Nanoemulsion to maintain wellbore stability in deep well drilling. Langmuir. 2022;**38**(41):12539-12550. DOI: 10.1021/acs.langmuir.2c01989

[58] Li Y, Wang M, Wei Z, An Y, Qin W, Bo K, et al. A novel oil-in-water nanoemulsion as a high-temperature plugging agent for wellbore wettability alteration and strengthening. Journal of Molecular Liquids. 2023;**384**:122273. Available from: https://www.sciencedirect.com/science/article/pii/S0167732223010760

Chapter 6

Nanoemulsions: A Recent Drug Delivery Tool

Vaibhav Changediya

Abstract

The use of nano/sub-micron particles in food, cosmetic, and pharmaceutical technology is becoming more and more popular. In particular, this interest has been growing in tandem with improved stabilization and emulsification methods. High-energy and low-energy spontaneous emulsification techniques are the two primary categories of nanoemulsion preparation techniques. Stability ranging from a few hours to years is influenced by important characteristics related to preparation procedures and components used. Nanoemulsions do not worry about issues like creaming, coalescence sedimentation, and flocculation because of their small droplet size. Ostwald ripening for them is the primary destabilizing process, though. This chapter provides a thorough overview of nanoemulsions including an explanation of their preparation techniques and assessments.

Keywords: nanoemulsion, spontaneous emulsification, Ostwald ripening, creaming, flocculation

1. Introduction

Recently, there has been a lot of interest in lipid-based formulations as a way to increase the permeability and bioavailability of poorly soluble drugs. As a result, many innovative drug delivery strategies have been used, and in each case, the nano-emulsion is essential to the active pharmaceutical ingredient's transport to the target organ or location. Among many different technologies, nanoemulsions have proven to advance drug delivery systems more than others.

These are considered the best alternatives for raising the oral bioavailability of drugs listed in the Biopharmaceutical Drug Classification System (BCS) under classes II and IV. A nanoemulsion is a clear solution of two nonsoluble liquids, such as water and oil, that is thermodynamically stable and stabilized by an interfacial layer of surfactant molecules. Nanoemulsions are a novel approach to medicine administration that uses emulsified water and oil systems with mean droplet sizes ranging from 50 to 1000 nm.

The main distinctions between emulsions and nanoemulsions are the size and morphology of the particles dispersed in continuous phases. The particle size in typical emulsions is 1/20 μm, but in nanoemulsions, it is 10/200 nm. A nanoemulsion is a kinetically stable liquid that consists of an oil phase and a water phase that possesses the appropriate surfactant. The dispersed phase has a low oil/water interfacial tension and is primarily composed of tiny particles with sizes ranging from 5 to 200 nm. They are regarded as the greatest alternative colloidal dispersions with exact ratios of an oil phase, an aqueous phase, a surfactant, and a co-surfactant are called nanoemulsions. To create the nanoemulsions, both high energy and low-energy emulsification procedures were applied. While low-energy emulsification methods take advantage of the system's physicochemical properties, which exploit phase transitions to produce nanoemulsion, high-energy emulsification methods utilize high shear mixing, high/pressure homogenization, or ultrasonification. Because oil, surfactant, and co-surfactant nanoemulsions are nontoxic and nonirritating, the Food and Drug Administration (FDA) has approved them for consumption by humans as "generally recognised as safe" [1, 2].

2. Composition of nanoemulsion

The following are the primary elements used in the creation of nanoemulsions:

1. Oil (for the solubilization of drugs or lipophilic molecules)

2. Surfactant

3. Water

4. Co-surfactant (amplify the surfactant's effect).

The stability of the system is greatly dependent upon and impacted by the selection of the appropriate components for nanoemulsions. Thus, choosing them carefully is essential. Surfactants and/or co-surfactants/co-solvents, along with an oily phase and an aqueous phase, make up most nanoemulsions.

2.1 Oil

Thus, oil is the most crucial ingredient in nanoemulsions. This facilitates the completion of the intended tasks. As such, it may operate as a carrier for the lipophilic active pharmaceutical ingredients or act as an active agent in and of itself (the essential oil in our study, for example). The selected oil's characteristics, including its polarity (low polarity oils are quickly disrupted by applied external force), viscosity (low viscosity oils are disrupted by applied external energy more quickly, resulting in the creation of tiny droplets more quickly), and interfacial tension (low interfacial tension in the oil facilitates size reduction and reduces the energy required to shrink the droplet size).

2.2 Surfactants

Without surfactants, the nanoemulsion system that stabilizes the thermodynamically unstable mixing of two immiscible liquids by lowering interfacial tension and altering dispersion entropy would be incomplete. The primary criteria for the surfactants used in the production of nanoemulsions are stability, high drug loading capacity, efficacious emulsification, and safety. Rapid adsorption of an appropriate surfactant used in the nanoemulsion formulation onto the interface between the two immiscible phases is necessary to prevent the coalescence of the nanodroplets and to dramatically lower interfacial tension. The four subgroups of surfactants are cationic, nonionic, zwitter-ionic, and anionic surfactants. Out of these four types, nonionic surfactants are used the most since they are known to have high biological absorption, are less sensitive to pH variations, and have a lower toxicity and irritating profile than ionic ones. The lipids in the stratum corneum can also be fluidized and solubilized by non-ionic surfactants, which enhances the skin's capacity to absorb the applied solution.

The scale used to classify nonionic surfactants is called the hydrophilic–lipophilic balance (HLB) which is a specific empirical expression of nonionic surfactants that characterizes the relationship between the hydrophilic and hydrophobic portions of the surfactant. In 1954, William C. Gryphon developed it. The values on the HLB scale range from 0 to 20. In order to form water/oil nanoemulsions, surfactants with HLB values between 3 and 6 are lipophilic (like SPANS) sorbitan monolaurate, whereas those with HLB values between 8 and 18 are hydrophilic (like TWEENS) polyoxyethylene sorbitan monolaurate. Co-surfactants are surfactants with an HLB value greater than 20. It is not a random addition of surfactants; rather, the surface becomes saturated with surfactants, and micelles begin to form when the critical micelle concentration is reached. It is claimed that a mixture of surfactants makes the nanoemulsion system more stable. The four subgroups of surfactants are cationic, nonionic, zwitter-ionic, and anionic surfactants. Nonionic surfactants are the most often used of these four types because they are less sensitive to pH changes and have a lower toxicity and irritating profile than ionic ones. They are also well known for having a high level of biological acceptability. The stratum corneum's lipids can also be fluidized and dissolved by nonionic surfactants.

They thereby enhance the skin's capacity to absorb and permeate the applied solution. William C. Gryphon developed a scale in 1954 that is used to group nonionic surfactants.

2.3 Co-surfactant

Co-surfactant helps the surfactant in the nanoemulsion system emulsify the oil in an aqueous phase. By combining with surfactants and penetrating the surfactant layer, co-surfactants in these systems dissolve the interfacial film, provide the necessary fluidity, reduce interfacial tension, and aid in the emulsification process. The interfacial film is made more flexible by the addition of a co-surfactant because surfactant alone frequently fails to produce temporary negative interfacial tension and fluid interfacial film.

Co-surfactants can also aid in the solubilization of the oil by altering the oil-water interface's curvature. Because the interaction between the surfactant and co-surfactant impacts how therapeutic compounds or lipophilic drugs partition into the aqueous and oil phases, choosing the right co-surfactant is crucial. 1,2-Propylene glycol, Transcutol® HP (diethylene glycol monoethyl ether), polyethylene glycol-400, and absolute carbohol. Alcohols may increase the miscibility of the two phases because of the way they partition between the aqueous and oily phases. Thus, ethanol, isopropyl alcohol, 1-butanol, and propylene glycol were selected as co-surfactants. Additionally, because carbosol and polyethylene glycol 400 are generally well-tolerated and show increased permeability when added to formulations, they were selected. The selection of surfactants and co-surfactants in the preparation of nanoemulsions is based on their % transmittance. Transcutol P (diethylene glycol monoethyl ether) had a higher % transmittance than propylene glycol. It was also found that transcutol P (diethylene glycol monoethyl ether) worked well as a potent permeation enhancer. It has been found that the surfactant's performance during the emulsification process is influenced by the co-surfactant concentration. Selecting a co-surfactant involves a number of factors, one of which is assessing the phase diagram's nanoemulsion area. The capacity of the ratio of surfactant to co-surfactant to form nanoemulsions is significantly influenced by the size of the nanoemulsion area in the phase diagram [3–5].

3. Methods of preparation of nanoemulsion

The stratum corneum (SC), the top layer of the epidermis, acts as a powerful barrier to the applied components, as was previously discussed. This issue makes it difficult for the formulations to penetrate the skin. Consequently, numerous solutions including micro and nanosystems were explored and developed in order to get around this problem. Adding the active substance or ingredients to a nanoemulsion carrier was one of the better ways. An emulsion with droplets ranging in size from 20 to 200 nm is called a nanoemulsion. It is a colloidal system with two phases that are incompatible with one another: an aqueous phase and an oily phase. The system's dispersion of the oily component into the aqueous phase is referred to as an oil-in-water (oil/water) nanoemulsion.

To maintain physical stability over the long term without any flocculation or coalescence in the system. A nanoemulsion is a transparent or translucent liquid-in-liquid dispersion system that is stable both thermodynamically and kinetically. Nanoemulsions offer an advantage over regular emulsions in that they are able to better encapsulate medications in the body. The ability of the nanometer-sized droplets to impede phase separation, coalescence, and flocculation improves the dispersibility of the system. Better long-term stability follows as a consequence [6, 7].

3.1 Phase inversion method

Chemical energy from phase changes that happen during the emulsification process produces fine dispersion. The required phase transitions are provided by changes in the polymer chain or composition at constant temperature. Predicated on the notion that temperature affects the solubility of polyoxyethylene/type surfactants. The surfactant monolayer exhibits a notable positive temperature change with constant composition at low temperatures. When temperatures rise, this surfactant becomes lipophilic and produces an oil-swollen micellar solution phase due to spontaneous curvature brought on by dehydration, as shown in **Figure 1**.

3.2 Phase inversion temperature (PIT)

This method changes the composition without changing the temperature. It is important to note that nonionic surfactants, including polyethoxylated surfactants, have temperature-dependent solubility. Emulsification is the process of changing the affinities of surfactants for water and oil that are dependent on temperature. Heat causes polyethoxylated surfactants to become lipophilic due to the dehydration of polyoxyethylene groups. As a result, this circumstance validates the applicability of the PIT method for producing nanoemulsions. To make nanoemulsions with this

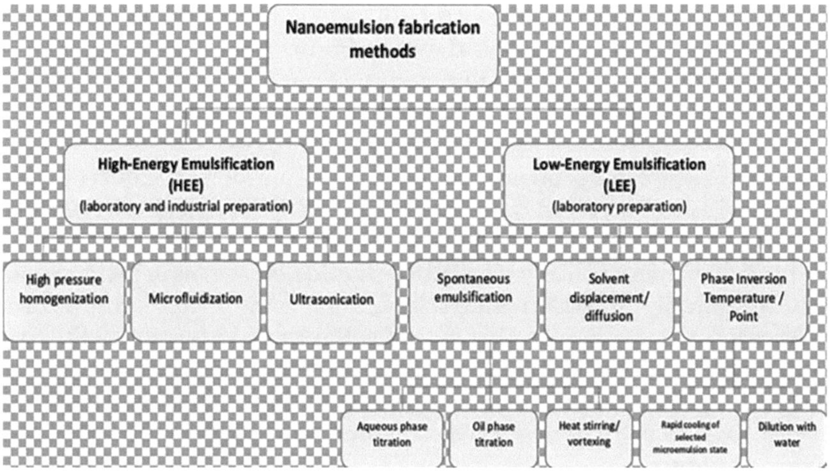

Figure 1.
Method of preparation of nanoemulsion.

method, the sample temperature must achieve the PIT (hydrophile–lipophile balance) or PIT level.

The PIT method produces the smallest droplet sizes and interfacial tensions that are feasible. This method improves emulsification by utilizing extraordinarily low interfacial tensions at the HLB temperature. It has been observed, therefore, that despite spontaneous emulsification at the HLB temperature, emulsions are incredibly delicate because of the incredibly fast coalescence rate. It has been reported that stable, fine emulsion droplets can be formed by rapidly cooling the emulsion to a temperature near PIT [8–10].

3.3 Phase inversion composition (PIC)

This method keeps the temperature constant while changing the composition. Nanoemulsions can be made by progressively adding water or oil to the water/surfactant or oil/surfactant mixture. The PIC technique is more appropriate for large-scale production than the PIT method since adding a single component to an emulsion is easier than producing a sudden change in temperature. When more water is supplied to the system, the water volume increases, and the system reaches a transition composition. As stated otherwise, the surfactant's spontaneous curvature goes from negative to zero as its polyoxyethylene chains get more hydrated. The transition composition achieves a balance between the lipophilic and hydrophilic properties of the surfactant, much like the HLB temperature, as shown in **Figure 1**.

3.4 Sonication method

Sonication is the most efficient way to make nanoemulsions. This method reduced the droplet size of conventional emulsions or microemulsions by using a sonication process. This method can only be used to generate small batches of nanoemulsions; large batches cannot be prepared this way. Even though ultrasound can create emulsion directly, it is best to create coarse emulsion before applying acoustic power since breaking an interface requires a lot of energy. Due to its low product throughput, the ultrasonic emulsification process is mostly utilized in laboratories to create emulsion droplets as thin as 0.2 micrometers, as shown in **Figure 1** [11].

3.5 Ultrasonic system

"Sonotrodes" (sonicator probes) are composed of piezoelectric quartz crystals that may expand and contract in response to alternating electrical voltage and provide the energy input for ultrasonic emulsification. Cavitations, which occur when the sonicator probe's tip mechanically vibrates the liquid, are the main process that produces ultrasonically generated effects. Cavitation is the formation and collapse of vapor cavities in a moving liquid. This type of vapor cavity arises when fluctuations in local velocity cause the local pressure to decrease to the temperature of the flowing liquid. When these cavities collapse, strong shock waves are produced that break up the scattered droplets by radiating across the solution along the radiating face of the tip, as shown in **Figure 1**.

3.6 Microfluidizer

When making emulsion, much higher pressures can be reached—up to about 700 Mpa. This is accomplished at the nozzle of the microfludizer, in the interaction chamber, where two jets of crude emulsion from different channels collide with one another. The process stream is delivered by a pneumatically powered pump that can pressurize the on-site compressed air (150/650 Mpa) to a maximum of 150 Mpa. A high-pressure flow stream driven via microchannels and into an impingement spot results in an impressive shearing motion. This has the potential to create a very fine emulsion.

3.7 High/energy emulsification method

Systems known as nanoemulsions are incapable of self-formation or equilibrium. This is why mechanical or chemical energy must be added to them in order for them to form. Among the mechanical energy input tools utilized in the creation of nanoemulsions are high-pressure homogenizers, high-shear stirring, and ultrasonic generators. These mechanical devices generate strong forces that separate the oil and water phases to form nanoemulsions. In high-energy technologies, the input energy density is approximately 108/1010 W kg/1. In the shortest length of time, the system obtains the energy required to generate uniform, small-sized particles. High/pressure homogenizers are the most widely used instruments for producing nanoemulsions because of their ability to achieve this.

3.8 High-pressure homogenizer

It is the method for producing nanoemulsions that is most frequently utilized. Nanoemulsions with particle sizes as small as 1 nm can be produced using this technology by employing a piston homogenizer or a high-pressure homogenizer. During the process, a small aperture is used to force the macroemulsion through at a pressure of 500–5000 psi. When cavitation, severe turbulence, and hydraulic shear come together, the process creates incredibly small droplet-sized nanoemulsions.

This process can be continued until the final product reaches the desired polydispersity index (PDI) and droplet size. The uniformity of droplet size in nanoemulsions is specified by PDI. Higher PDI in nanoemulsions is connected with lower droplet size uniformity. A PDI of less than 0.08 indicates a monodisperse sample; a PDI of more than 0.3 indicates a narrow size distribution; and a PDI between 0.08 and 0.3 indicates a broad size distribution. However, the production of small droplets at the submicron level requires a large amount of energy. This energy combined with the increasing temperatures during the high-pressure homogenization process, could cause the components to degrade. Enzymes, proteins, and nucleic acids are a few examples of molecules that heat can damage.

3.9 High/shear stirring

This method uses high-energy mixers and rotor/stator systems to prepare nanoemulsions. The droplet sizes of the internal phase can be significantly decreased by increasing the mixing intensity of these devices. However, creating emulsions with an average droplet size of less than 200–300 nm could be difficult [12–16].

3.10 Low/energy emulsification method

Nanoemulsification can also be accomplished with low-energy methods, resulting in more homogeneous and smaller droplets. These methods such as phase inversion temperature and phase inversion component, produce smaller and more homogeneous droplets by making use of the physicochemical properties of the system. The types of oils and emulsifiers that low-energy techniques can employ are restricted, such as proteins and polysaccharides, even though they are frequently more effective at creating minute droplets than high-energy approaches. To solve this problem, large amounts of artificial surfactants are used in low-energy methods to produce nanoemulsions; nonetheless, this limits their range of uses, especially for many food processing as shown in **Figure 1**.

3.11 Spontaneous nanoemulsification

It benefits from the chemical energy replacement during the emulsification process, which is based on dilution with the continuous phase and usually proceeds at constant temperature without any phase transitions in the system. This method can produce nanoemulsions at room temperature, hence no specialized equipment is required. Essentially, the following variables influenced it: bulk and interfacial viscosity, phase transition region, surfactant concentration, surfactant structure, and interfacial tension. In the pharmaceutical industry, systems developed with this method are usually called self-emulsifying drug/delivery systems (SEDDS) or self-nano/emulsifying drug/delivery systems (SNEDDS). When water is mixed with an oil phase that has a water-soluble component, oil droplets naturally form. The process is driven by the movement of a chemical that dissolves in water from the oil phase to the water phase. This leads to interfacial turbulence, which helps oil droplets form naturally, as shown in **Figure 1** [17–19].

4. Patents related to nanoemulsion

As the most effective type of intellectual property protection, patents are crucial to a nanotechnology company's expansion. Patents will be crucial to the success of the global nanotechnology revolution, just as they were to the growth of the biotechnology and information technology industries. In fact, patents are already influencing the young, quickly developing field of nanoscience and small technologies. A company's long-term survival will depend on its ability to get legitimate and defendable patent protection as it develops nanotechnology-related products and processes and starts looking for commercial uses for its ideas, as stated in **Table 1** [20–49].

5. Evaluation of nanoemulsion

5.1 Determination of encapsulation efficiency

To determine the amount of drug entrapped in the formulation, a weighed quantity of the formulation is ultrasonically dispersed in an organic solvent to release the drug, which is then extracted into a suitable buffer. Once the right dilutions are made and compared to an adequate blank, the extraction is subjected to spectrophotometric analysis at the drug's λ_{max} to determine the drug content. The drug's loading efficiency (LE) and entrapment efficiency (EE) can be calculated using these formulas. Drug LE represents the amount of drug in the achieved product (mg)/ total product weight (mg) × 100, while drug EE represents the amount of drug in the obtained product (mg)/total quantity of drug added (mg) × 100. To assess drug concentration, high-performance liquid chromatography (HPLC) in reverse phase can also be employed [50–57].

5.2 Determination of particle size and polydispersity index (PDI)

Photon correlation spectroscopy (PCS) is used to measure the PDI and particle size of nanoemulsions using a Malvern Zetasizer. PCS calculates the variation in light scattering over time brought on by particle Brownian motion.

The fundamental tenet of PCS is that smaller particles accelerate larger ones. The laser beam is distorted by the submicron particles present in the solution. The rate at which particle diffusion causes the laser scattering intensity to vary around a mean value at a fixed angle depends on the particle size. The calculated photoelectron time correlation function yields a line width distribution histogram that is associated with particle size. A weighed quantity of formulation is combined with double or distilled water to generate a homogenous dispersion, which is then used to measure the particle size. The PDI and particle size must be measured using this mixture immediately. A 0 (zero) PDI represents a monodisperse system while a 1 PDI represents a polydisperse particle dispersion [20].

5.3 Determination of zeta potential

The zeta potential is a method for figuring out a particle's surface charge in a liquid. Zeta potential is a helpful tool for predicting dispersion stability; the presence and adsorption of electrolytes, as well as the physicochemical properties of the drug, polymer, and medium, all affect its value. The Malvern Zetasizer apparatus is employed for its measurement. Zeta potential is determined by diluting the nanoemulsion and calculating its value using the electrophoretic mobility of the oil

Patent application title	Patent App. No.	Date
Nanoparticles and nanoemulsions	14/893,123	2017/12/05
Nanoemulsion therapeutic compositions and methods of using the same	12/567571	2015/02/24
Antimicrobial nanoemulsion compositions and methods	8236335	2012/08/07
Antimicrobial nanoemulsion compositions and methods	8232320	2012/07/31
Topical compositions and methods of detection and treatment	20120039814	2012/02/16
Cancer vaccine compositions and methods of using the same	20110280911	2011/11/17
Methods of using nanoemulsion compositions having anti/inflammatory activity	20110200657	2011/08/18
Stable nanoemulsions for ultrasound/mediated of drug delivery and imaging	20110177005	2011/07/21
Method for the preparation of nanoparticles from nanoemulsion	20110135734	2011/06/09
Antimicrobial nanoemulsion compositions and methods	20110070306	2011/03/24
Nanoemulsion formulations for direct delivery	20110045050	2011/02/24
Lyophilized nanoemulsion	20110015266	2011/01/20
Nanoemulsion vaccines	20100316673	2010/12/16
Nanoemulsion of resveratrol/phospholipid complex and method for preparing the same and applications thereof	20100297199	2010/11/25
Per fluorocarbonnanoemulsion containing quantum dot nanoparticles and method for preparing the same	20100233094	2010/09/16
Compositions for treatment and prevention of acne, methods for making the compositions, and methods of use thereof	20100226983	2010/09/09
Nanoemulsion therapeutic compositions and methods of using the same	20100092526	2010/04/15
Stable mixed emulsions	20100069511	2010/03/18
Antimicrobial nanoemulsion compositions and methods	7655252	2010/02/02
Antimicrobial nanoemulsion compositions and methods	20100003330	2010/01/07
Nanoemulsion influenza vaccine	20090304799	2009/12/10
Nanoemulsion adjuvants	20090291095	2009/11/26
Oil/in/water nanoemulsion, a cosmetic composition, and a cosmetic product comprising it, a process for preparing said nanoemulsion	20090208541	2009/08/20
Nanoemulsion therapeutic compositions and methods of using the same	20080317799	2008/12/25
	PCT/AU2008/001714	2008/11/18
Nanoemulsion vaccines	20080254066	2008/10/16
Nanoemulsion vaccines	20080181905	2008/07/31
Nanoemulsion vaccines	7314624	2008/01/01
Nanoemulsion compositions having anti/inflammatory activity	20070036831	2007/02/15
Nanoemulsion formulations	20020155084	2002/10/24
Nanomulsion based on nonionic and cationic amphiphilic lipids and uses thereof	6039936	2000/03/21

Patent application title	Patent App. No.	Date
Solid fat nanoemulsions as vaccine delivery vehicles	5716637	1998/02/10
Solid fat nanoemulsions as drug delivery vehicles	5576016	1996/11/19

Table 1.
Patents on nanoemulsion.

droplets. It is believed that a zeta potential of ±30 mV suffices to ensure the nanoemulsion's physical stability.

5.4 Morphological study of nanoemulsion

The morphology of nanoemulsions is analyzed using transmission electron microscopy (TEM). In a transmission electron microscope (TEM), an electron beam is directed at a thin foil specimen. Upon interaction with the material, these incident electrons become unscattered, elastically scattered, or elastically scattered electrons. The distances between the objective lens and the specimen and between the objective lens and its image plane regulate the magnification.

The electromagnetic lenses focused the scattered or unscattered electrons and projected them onto a screen to produce a contrast/amplitude image, a phase/contrast image, an electron diffraction image, or a phantom image with distinctly darkened parts, depending on the density of unscattered electrons. At higher magnifications, diffraction modes can be used in conjunction with bright field imaging to reveal the size and structure of nanoemulsion droplets. To immobilize the sample for TEM investigation, a suspension of lyophilized nanoparticles or a few drops of nanoemulsion are prepared in double-distilled water and placed onto a holey film grid. The excess solution must be squeezed off the grid and dyed after immobilization [21, 22].

5.5 Atomic force microscope (AFM)

These days, the surface morphology of nanoemulsion formulations is studied using a relatively new technique called AFM. Nanoemulsions are diluted with water and then applied to a glass slide in order to conduct AFM. After that, the coated drops are dried in an oven and scanned at 100 mV/s.

5.6 *In vitro* drug release study

Performance of drug formulation drug release studies conducted *in vitro* can help estimate *in vivo*. The *in vitro* release rate of a medication is usually studied using a United state pharmacopoeia (USP) dissolving apparatus. Dried nanoparticles or nanoemulsion containing 10 mg of medication were applied to dialysis membrane pouches and placed in a flask containing buffer after being dissolved in buffer. This experiment is run at 37 ± 0.5° with a stirring speed of 50 rpm. Samples are removed on a regular basis and are always replaced with the same volume of fresh dissolving media.

Spectrophotometry is used to detect the absorbance of materials at a certain wavelength after they have been suitably diluted. The absorbance of the sample is used to compute the percentage of drug release at different time intervals using a calibration curve.

5.7 *In vitro* skin permeation studies

With the Keshary–Chien/diffusion cell, permeation investigations can be studied both *in vitro* and *in vivo*. For permeation studies, the abdominal skin of adult male rats weighing 250 ± 10 g is frequently utilized. The donor and recipient chambers of the diffusion cell are separated by the rat skin. Twenty percent ethanol-infused fresh water is added to receiver chambers that are kept at 37°C and rotated at 300 rpm all the time. The formulas are kept in the donor room.

At predetermined intervals, such as 2, 4, 6, and 8 hours, a predetermined volume (0.5 ml) of the receiver chamber's solution was removed for gas chromatographic analysis. Each time, the sample was quickly replaced with an equivalent volume of brand-new solution. Three runs are made for each sample. The total amount of medication absorbed via rat skins at each time interval is plotted against a function of time following cumulative adjustments. In steady-state conditions, the plot slope is utilized to calculate medicine penetration rates [23].

5.8 Stability studies

The purpose of stability studies is to assess how stable a medicinal ingredient is in the presence of various environmental factors, including temperature, humidity, and light. The formulation is stored for 24 months in a dispersed and freeze/dried state before the stability studies of the nanoemulsion are carried out in accordance with the guidelines set forth by the International Conference on Harmonization. There are three storage conditions that are utilized: ambient (25 ± 2°/60 ± 5% RH), refrigerated (5 ± 3°), and freeze (−20 ± 5°). The necessary quantity of nanoemulsion is stored in glass vials with hermetically sealed closures. Samples are removed and analyzed for characteristics such as loading, particle size, EE, and the *in vitro* drug release profile at predetermined intervals [24].

5.9 Shelf life determination

Expedited stability studies are performed to determine the shelf life of a nanoemulsion. The formulations are stored at three distinct temperatures and relative humidity levels (30°, 40°, and 50 ± 0.5°) for almost 3 months. After a predetermined period of time (0, 30, 60, and 90 days), samples are taken out and analyzed using HPLC at λmax to find out how much medication remains. The samples that are removed at zero time are known as control samples. This determines the order of the reaction. Next, using the following equation, the reaction rate constant (K) for the deterioration is calculated at each elevated temperature based on the slope of the lines: An Arrhenius plot with slope = −K/2.303 is produced by plotting the logarithms of K at different elevated temperatures against the reciprocal of absolute temperature. This data yields the plot value of K at 25°, which is then used to calculate shelf life by plugging the value into the calculation $t_{0.9} = 0.1052/K_{25}$. Here, $t_{0.9}$—also referred to as shelf life—is the amount of time required for a 10% degradation of the drug [25].

5.10 Thermodynamic stability studies

The majority of thermodynamic stability studies are carried out in three steps. Initially, a cycle of heating and cooling is conducted to determine whether adjusting the temperature affects the stability of the nanoemulsion in any way. The nanoemulsion is subjected to six cycles of temperature between 4° (the refrigeration temperature) and 40° while the formulation is stored at each temperature for a minimum of 48 hours.

The formulations that show stability at these temperatures are used in centrifugation experiments. The second technique involves centrifuging the manufactured nanoemulsions for 30 minutes at 5000 rpm in order to check for phase separation, creaming, or cracking. Those who showed no signs of instability could experience a freeze-thaw cycle. The third technique is the freeze/thaw cycle, which entails freezing and thawing nanoemulsion formulations three times at temperatures between –21° and + 25°.

If a formulation shows no signs of instability, this test finds that it has good stability. These mixtures are then subjected to dispersibility tests to determine how well they self- or emulsify. This test determines that a formulation has good stability if it exhibits no indicators of instability. The next step is to put these mixtures through dispersibility tests to see how effectively they self- or emulsify.

5.11 Dispersibility studies

Dispersibility tests are performed utilizing conventional USP XXII dissolving equipment to assess the effectiveness of self-emulsification or nanoemulsion formation. Each formulation is added to 500 ml of distilled water that is kept at 37 ± 0.5°; 2.1 ml of each formulation is used. A typical dissolution paddle made of stainless steel spins at 50 revolutions per minute to gently agitate the mixture. The nanoemulsion formulations' *In vitro* performance is assessed visually by the application of a grading system that is explained below. Level A nanoemulsions seem clear or bluish and form quickly—within 1 minute. Although they are somewhat less transparent, Grade B nanoemulsions develop quickly and have a bluish-white appearance. Within 2 minutes, grade C nanoemulsions, which are fine, milky emulsions form. Grade D emulsions are slower and more dull, with a grayish-white look and a hint of oiliness [26].

5.12 Determination of viscosity

Viscosity measurement is a crucial component of a nanoemulsion's physicochemical assessment. Viscosity is measured using a number of tools, such as the Ostwald viscometer, Hoeppler falling ball viscometer, Stormer viscometer, Brookfield viscometer, and Ferranti/Shirley viscometer. Of all these devices, the Brookfield viscometer is the one that is suggested for figuring out the viscosity of a nanoemulsion.

The assessment of viscosity verifies whether the system is an O/W or W/O emulsion. Systems that exhibit low viscosity are classified as O/W types, whereas those that exhibit high viscosity are classified as water-in-oil types. Nonetheless, the survismeter has emerged as the most often utilized piece of apparatus since it assesses the particle size, hydrodynamic volumes, contact angle, dipole moment, interfacial tension, surface tension, and viscosity of the nanoemulsions.

5.13 Refractive index

The refractive index provides information on the transparency of the nanoemulsion and the transmission of light through the medium. The ratio of the wave's phase speed (vp) in the medium to its speed (c) in the reference medium is the formula for determining a medium's refractive index (n), and it is expressed as $n = c/vp$. The refractive index of the nanoemulsion can be determined by placing a drop on a slide and comparing it to the refractive index of water (1.333) using an Abbes-type refractometer set at 25 ± 0.5. In order for a nanoemulsion to be considered transparent, its refractive index must equal that of water.

5.14 pH and osmolarity measurements

A pH meter is used to determine a nanoemulsion's pH, whereas a microosmometer uses the freezing point method to determine an emulsion's osmolarity. Once 100 μl of nanoemulsion has been transferred into a microtube, measurements are taken.

5.15 Dye solubilization

A water soluble dye is dispersible in an O/W globule as opposed to soluble in the aqueous phase of the W/O globule. Similarly, an oil-soluble dye is dispersible in the W/O globule but dissolves in the oily phase of the O/W globule. An O/W nanoemulsion will absorb color evenly when water-soluble dye is applied; however, if the emulsion is W/O, the dye will only remain in the dispersed phase, and the color will not absorb evenly [27].

5.16 Dilutability test

The stability of a nanoemulsion can be preserved when a continuous phase is added in larger quantities which justifies the dilution test. Consequently, unlike O/W nanoemulsions, W/O nanoemulsions do not dilute with water and instead experience phase inversion into O/W nanoemulsions. W/O nanoemulsion can only be diluted with oil [28].

Conflict of interest

The authors declare no conflict of interest.

Abbreviations

BCS	biopharmaceutical drug classification system
HLB	hydrophilic-lipophilic balance
SC	stratum corneum
PIT	phase inversion temperature
PIC	phase inversion composition
PDI	polydispersity index
SEDDS	self-emulsifying drug/delivery systems
SNEDDS	self-nano/emulsifying drug/delivery systems
LE	loading efficiency
EE	entrapment efficiency
HPLC	high-performance liquid chromatography
PCS	Photon correlation spectroscopy
TEM	transmission electron microscopy
AFM	atomic force microscope

Author details

Vaibhav Changediya
Dr. Vedprakash Patil Pharmacy College, Aurangabad, Maharashtra, India

*Address all correspondence to: vchangediya@gmail.com

References

[1] Jaiswal M, Dudhe R. Nanoemulsion: An advanced mode of drug deliverysystem. Biotech. 2015;**5**:123-127. DOI: 2010.1007/s13205-014-0214-0

[2] Dorkoosh FA, Brussee J, Verhoef JC, Borchard G, Rafiee Tehrani M, Junginger HE. Preparation and NMR characterization of superporous hydrogels and Superporous hydrogel composites. Polymer. 2000;**41**(23):8213-8220. DOI: 2010.1016/S0032-3861(00)00200-7

[3] Okyar A, Özsoy Y, GüngÖr S. Novel formulation approaches for dermal and transdermal delivery of non-steroidal anti-inflammatory drugs. In: Lemmey A, editor. Rheumatoid Arthritis–Treatment. Croatia: Intech; 2012. pp. 25-48. DOI: 10.5772/28461

[4] Sunil KY, Manoj KM, Anumapaa T, Ashutosh S. Emulgel a new approach for enhanced for topical drug delivery. International Journal of Current Pharmaceutical Research. 2017;**9**(1): 15-19. DOI: 10.22159/ijcpr.16628

[5] Hardenia A, Jayronia S, Jain S. Emulgel: An emergent tool in topical drug delivery. International Journal of Pharmaceutical Sciences Review and Research. 2014;**5**(5):1653-1660. DOI: 2010.13040/IJPSR.0975-8232.5(5).1653-60

[6] Jana S, Ali SA, Nayak AK, Sen KK, Basu SK. Development of topical gel containing aceclofenac-crospovidone solid dispersion by "quality by design (QbD)" approach. Chemical Engineering Research and Design. 2014;**9**(2):2095-2105. DOI: 10.1016/j.cherd.2014.01.025

[7] Mohamed MI. Optimization of chlorphenesin emulgel formulation. The AAPS Journal. 2004;**6**(3):81-87. DOI: 2010.1208/aapsj060326

[8] Jones DS, Woolfson AD, Brown AF. Textural, viscoelastic and mucoadhesive properties of pharmaceutical gels composed of cellulose polymers. International Journal of Pharmaceutics. 1997;**151**(2):223-233. DOI: 10.10 16/S0378-5173(97)04904-1

[9] Andrews GP, Gorman SP, Jones DS. Rheological characterization of primary and binary interactive bioadhesive gels composed of cellulose derivatives designed as ophthalmic viscosurgical devices. Biomaterials. 2015;**26**(5):571-580. DOI: 2010.1016/j.biomaterials.2004.02.062

[10] Rai VK, Mishra N, Yadav KS, Yadav NP. Nanoemulsion as pharmaceutical carrier for dermal and transdermal drug delivery: Formulation development, stability issues, basic considerations and applications. Journal of Controlled Release. 2018;**270**:203-225. DOI: 2010.1016/ j.jconrel.2017.11.049

[11] John J, Bhattacharya M, Raynor PC. Emulsions containing vegetable oils for cutting fluid application. Colloids and Surfaces A: Physicochemical and Engineering Aspects. 2004;**237**(1):141-150. DOI: 2010.1016/j.colsurfa.2003.12.029

[12] Mehmood T, Ahmad A, Ahmed A, Ahmed Z. Optimization of olive oil based O/W nanoemulsions prepared through ultrasonic homogenization: A response surface methodology approach. Food Chemistry. 2017;**229**:790-796. DOI: 2010.1016/j.foodchem.2017.03.023

[13] Lu W-C, Chiang B-H, Huang D-W, Li P-H. Skin permeation of

d-limonene-based nanoemulsions as a transdermal carrier prepared by ultrasonic emulsification. Ultrasonics Sonochemistry. 2014;**21**(2):826-832. DOI: 2010.1016/j.ultsonch.2013.10.013

[14] Jafari SM, He Y, Bhandari B. Nano-emulsion production by sonication and microfluidization-a comparison. International Journal of Food Properties. 2006;**9**(3):475-485. DOI: 2010.1080/10942910600596464

[15] Saberi AH, Fang Y, McClements DJ. Effect of glycerol on formation, stability, and properties of vitamin-E enriched nanoemulsions produced using spontaneous emulsification. Journal of Colloid and Interface Science. 2013;**411**:105-113. DOI: 2010.1016/j.jcis.2013.08.041

[16] Sriamornsak P, Limmatvapirat S, Piriyaprasarth S, Mansukmanee P, Huang Z. A new self-emulsifying formulation of mefenamic acid with enhanced drug dissolution. Asian Journal of Pharmaceutical Sciences. 2015;**10**:121-127. DOI: 10.1016/j.ajps.2014.10.003

[17] Patel KC, Pramanik S. Formulation and characterization of mefenamic acid loaded polymeric nanoparticles. World Journal of Pharmacy and Pharmaceutical Sciences. 2014;**3**(6):1391-1405. 2097348896

[18] Oswal T, Naik S. Formulation and evaluation of mefenamic acid emulgel. IJPRD. 2014;**5**(12):91-100. DOI: 10.1016/j.jsps.2011.08.001

[19] Kamboj S, Saini V, Bala S, Sharma G. Formulation and characterization of drug loaded niosomal gel for anti-inflammatory activity. International Journal of Medical, Health, Biomedical, Bioengineering and Pharmaceutical Engineering. 2013;7(2):541-545. 207231875

[20] Nanoparticles and nanoemulsions 14/893,123. 2017

[21] Nanoemulsion therapeutic compositions and methods of using the same 12/567571. 2015

[22] Antimicrobial nanoemulsion compositions and methods 8236335. 2012

[23] Antimicrobial nanoemulsion compositions and methods 8232320. 2012

[24] Topical compositions and methods of detection and treatment 20120039814. 2012

[25] Cancer vaccine compositions and methods of using the same 20110280911. 2011

[26] Methods of using nanoemulsion compositions having anti/inflammatory activity 20110200657. 2011

[27] Stable nanoemulsions for ultrasound/mediated of drug delivery and imaging 20110177005. 2011

[28] Method for the preparation of nanoparticles from nanoemulsion 20110135734. 2011

[29] Antimicrobial nanoemulsion compositions and methods 20110070306. 2011

[30] Nanoemulsion formulations for direct delivery 20110045050. 2011

[31] Lyophilized nanoemulsion 20110015266. 2011

[32] Nanoemulsion vaccines 20100316673. 2010

[33] Nanoemulsion of resveratrol/phospholipid complex and method for preparing the same and applications thereof 20100297199. 2010

[34] Per fluorocarbon nanoemulsion containing quantum dot nanoparticles and method for preparing the same 20100233094. 2010

[35] Compositions for treatment and prevention of acne, methods for making the compositions, and methods of use thereof 20100226983. 2010

[36] Nanoemulsion therapeutic compositions and methods of using the same 20100092526. 2010

[37] Stable mixed emulsions 20100069511. 2010

[38] Antimicrobial nanoemulsion compositions and methods 7655252. 2010

[39] Antimicrobial nanoemulsion compositions and methods 20100003330. 2010

[40] Nanoemulsion influenza vaccine 20090304799. 2009

[41] Nanoemulsion adjuvants 20090291095. 2009

[42] Oil/in/water nanoemulsion, a cosmetic composition and a cosmetic product comprising it, a process for preparing said nanoemulsion 20090208541. 2009

[43] Nanoemulsion therapeutic compositions and methods of using the same 20080317799. 2008

[44] Nanoemulsion vaccines 20080254066. 2008

[45] Nanoemulsion vaccines 20080181905. 2008

[46] Nanoemulsion vaccines 7314624. 2008

[47] Nanoemulsion compositions having anti/inflammatory activity 20070036831. 2000

[48] Solid fat nanoemulsions as vaccine delivery vehicles 5716637. 1998

[49] Solid fat nanoemulsions as drug delivery vehicles 5576016. 1996

[50] Faiyaz S, Shafiq S, Sushma T, Farhan J, Khar R, Ali M. Development and bioavailability assessment of ramipril nanoemulsion formulation. European Journal of Pharmaceutics and Biopharmaceutics. 2007;**66**:227-243

[51] NagibA. Nanoemulsion and Nanoemulgel as a TopicalFormulation. IOSR Journal of Pharmacy. 2015;**5**(10):43-47. DOI: 283046065

[52] Funk JL, Frye JB, Oyarzo JN, Chen J, Zhang H, Timmermann BN. Anti-inflammatory effects of the essential oils of ginger (*Zingiber officinale* roscoe) in experimental rheumatoid arthritis. Pharma Nutrition. 2016;**4**(3):123-131. DOI: 10.1016/j.phanu.2016.02.004

[53] Pavoni L, Perinelli DR, Bonacucina G, Cespi M, Palmieri GF. An overview of micro- and nanoemulsions as vehicles for essential oils: Formulation, preparation, and stability. Nanomaterials. 2020;**10**(1):135. DOI: 10.3390/nano10010135

[54] Salim N, Basri M, Rahman MA, Abdullah DK, Basri H, Salleh AB. Phase behaviour, formation and characterization of palm-based esters nanoemulsion formulation containing ibuprofen. Journal of Nanomedicine and Nanotechnology. 2011;**2**(4):1-5. DOI: 10.4172/2157-7439.1000113

[55] Chuacharoen T, Prasongsuk S, Sabliov CM. Effects of surfactant concentrations on physicochemical properties and functionality curcumin nanoemulsions under conditions relevant to commercial utilization. Molecules.

2019;**24**:2744. DOI: 10.3390/molecules 24152744

[56] Sakulku U, Nuchuchua O, Uawongyart N, Puttipipatkhachorn S, Soottitantawat A, Ruktanonchai U. Characterization and mosquito repellent activity of citronella oil nanoemulsion. International Journal of Pharmaceutics. 2009;**372**(1):105-111. DOI: 1016/j. ijpharm.2008.12.029

[57] Thakkar HP, Khunt A, Dhande RD, Patel AA. Formulation and evaluation of itraconazole nanoemulsion for enhanced oral bioavailability. Journal of Microencapsulation. 2015;**32**(6):559-569. DOI: 10.3109/02652048.2015.1065917

Chapter 7

Polymeric Nano-Emulsion in Functional Textile Finishing

Sana Javaid, Nadia Saleem and Shafi Ur Rehman

Abstract

Advancement in nanotechnology brings a revolutionary change in the field of textile finishing. Textile finishing is a chemical or a mechanical process to impart functional properties to the textile to provide comfort for wearer. Today's textile manufacturers focus on the manufacture of smart and functional textiles that are equipped with antifouling, anti-wrinkle, crease-resistant, water-repellent, flame-retardant, and soil-repellent properties for consumers' safety and well-being. A wide variety of functional chemical finishes are available in the market to meet the ongoing challenges in the textile sector. Nano-emulsions significantly contribute to a wide variety of functional finishes to provide advanced hi-tech applications for present and future textile consumers. Both natural and synthetic polymers have been utilized for the synthesis of functional finishes by employing polymeric nano-emulsions on cotton, wool polyester fiber as well as textile. Thus, nano-emulsions provide an inherent property to textile and stimulate the economic growth of functional textile market.

Keywords: polymeric nano-emulsions, functional textile finishing, polymeric nanoparticles, polymeric formulations, biomedical and industrial applications

1. Introduction

Emulsions comprise at least two immiscible liquids, presenting a dispersed phase within a continuous phase. They are classified as microemulsions, nano-emulsions, or macroemulsions, based on droplet size, appearance, and stability. Both macro-emulsions and nano-emulsions demonstrate thermodynamic instability [1]. In the late 1990s, various research groups introduced the term "nano-emulsion," which gained prevalence by the early 2000s coinciding with the declining use of the term "mini-emulsion" [2]. The term "nano-emulsion" was initially coined by Calvo in a 1996 peer-reviewed journal article, referring to colloidal systems with droplet sizes ranging from 200 to 250 nm [3]. Nano-emulsions denote the dispersion of minuscule droplets, at the nanometer scale, of one immiscible liquid within another immiscible liquid system, commonly oil and water, achieved through the assistance of surfactants [4]. Nano-emulsions are extensively researched within the field of pharmaceutics for their capacity to improve aqueous solubility [5–7]. These are characterized by their ultrafine emulsion state, featuring droplets measuring less than 100 nm in diameter. Unlike microemulsions, nano-emulsions exhibit metastable behavior, wherein their structural configuration is contingent upon the specific preparation methodology employed [8]. Presently, the generally accepted range for defining nano-emulsion systems predominantly spans droplet sizes of 20–200 nm [9]. In recent years, research emphasis has centered on formulating nano-emulsions through distinct methodologies, broadly categorized as high-energy and low-energy methods [10]. Mechanical equipment is required when using a high-energy approach [11]. The low-energy technique harnesses the intrinsic system's own energy, leveraging alterations in surfactant layer's spontaneous curvature around the emulsion droplets [12, 13].

The usage of nanomaterials and nanotechnology-based processes is expanding rapidly across all disciplines of research and technology. The textile sector is likewise reaping the benefits of nanotechnology in its vast range of applications. Textile-based products ranging from nanocomposite fibers, nanofibers as well as polymeric nanocoatings are used to impart functional and performance properties to traditional textile as well as for high-performance advanced textile applications. Nanoparticle use during standard textile processing, such as finishing, coating, and dyeing, significantly improves product performance and provides multiple functionalities [14].

Surface treatments can often modify regular textile materials with functional finishes for diverse applications, such as antibacterial, flame-retardant, and dirt release effects. Functional textiles are described as a subclass of technical textiles that provide specialized functional features [15]. Functional textiles are currently widely used in a range of industrial and biomedical applications, such as fashion, pharmaceutical, medical, and engineering.

There has been significant progress in natural and synthetic textile finishing methods, smart fabrics, and high-performance functional textiles [16]. Textiles or textile structures are widely used in high-performance technological fields, including furniture, food packaging, protective textiles, automotive, aerospace, and medical textiles [17]. The textile sector represents a major segment in the realm of nanotechnology research that stands as an early adopter to effectively showcase the practical applications of nanotechnology for consumers. Textile-focused research centers on leveraging nanoscale substances and creating nanostructures in manufacturing and finishing processes [14].

Nanotechnology overcomes the limitations of existing methods for imparting special properties to textiles [18]. The unique properties of synthetic textiles, such as self-cleaning, water- and oil-repellent, antistatic, antibacterial, stainproof, ultraviolet (UV) shielding, and moisture control, make them ideal for everyday use without sacrificing the fabric's toughness and breathability [19]. The growing desire for long-lasting, visually appealing, and operationally superior textile goods, along with a few sustainability issues, has enabled scientists to apply nanotechnology to the textile industry [20]. Thus, the world economy greatly relies on textile industry and textile manufacturers now focusing on the current aspects of growing needs of textile consumers toward health, hygiene, technology, and biomedical significance. Various functional treatment and methods have been utilized for the development of functional textile and textile materials [21].

Various surface modification techniques, such as electrospinning, plasma treatment, polymerization, microencapsulation, and sol-gel methodologies, have previously been employed to confer new functional attributes like water repellency, flame resistance, and antibacterial properties to textiles [22–24]. Thus, this chapter explores various types of functional textiles based on polymeric nano-emulsions in enhancing the inherent properties of the textile substrate. The role of nano-emulsions for imparting scalable, controlled, and multiple functional properties to textiles is gaining significant attraction among polymer and textile scientists and

Figure 1.
Types of functional textile based on nano-emulsions.

manufacturers worldwide. Polymeric nano-emulsion via slow and sustained release of active moieties to the textile fabric either woven or nonwoven greatly enhanced the potential applications toward industry worldwide. From antimicrobial performance to moisture management and ultraviolet (UV) protection, polymeric nano-emulsions serve as versatile carriers, revolutionizing functional textile, as shown in **Figure 1**.

2. Types of functional textiles

2.1 Polymeric nano-emulsion in antimicrobial textile finishing

Antimicrobial textiles refer to a specially designed functional textile or textile fabric designed to kill or repel the microorganisms [25]. Antimicrobial textiles are classified according to their antimicrobial activity, such as antibacterial, antifungal, or antiviral, as shown in **Figure 2**. Several antimicrobial textiles can protect against germs, fungi, and viruses all at the same time. Antimicrobial chemicals can target a wide range of microorganisms [26]. In impoverished nations, pathogenic bacterial infections caused by an improper handling of medical equipment and textiles lead to increase in morbidity rates and significant deaths [27]. There is a substantial demand for antimicrobial textiles in public settings, such as hotels, restaurants, and transportation systems, due to concerns about potential infection sources like towels, curtains, and carpets. These textiles can harbor various microorganisms that pose a risk of transmission between individuals. Continuous laundering effectively reduces microbial presence on fabrics; however, this is unfeasible in settings like hospitals with constant shifts. To address this aspect, the development of antimicrobial textiles is crucial, particularly for individuals working in sanitation and high-risk environments such as sewage treatment [28].

Bacterial infection in textile consumers is becoming more severe in moist and humid environments that facilitate the abrupt growth and multiplication of microorganisms [29]. To prevent bacterial growth and multiplication, it is necessary to

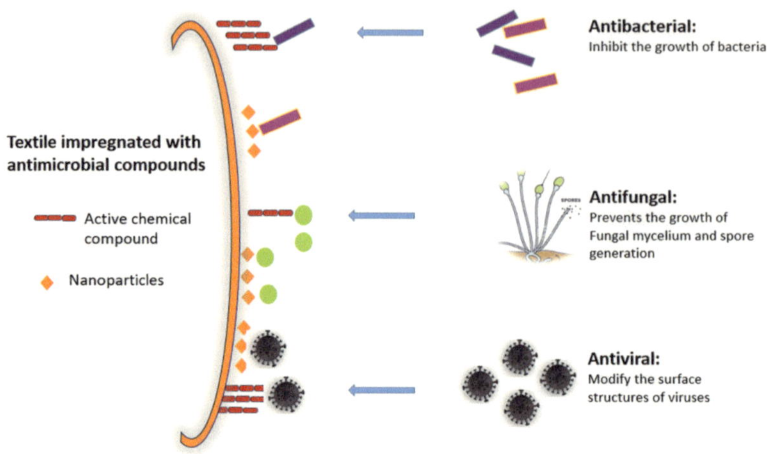

Figure 2.
Types of functional textile finishing.

destroy their cell membrane by destroying their protein structure and DNA [30]. Textiles undergo antibacterial treatment to prevent the growth of microorganisms by using biocidal chemical agents that cause the denaturization of bacterial colonies [31].

Antimicrobial textiles find applications in diverse domains, such as first aid, clinical, and hygienic practices. Chitosan-based polymeric nano-emulsions in encapsulating various active reagents as core have been employed to impart bi- or trifunctional properties to the textile as antimicrobial, antioxidant, and antifungal [32]. These materials are also employed to convert advanced pharmaceutical nanocarriers into conventional textiles, establishing wearable medication delivery methods. These bifunctional textiles represent favorable innovations capable of enhancing dermal penetration while minimizing the associated toxicity risks [33]. Textile functionalization with antifouling chemicals has enabled several biomedical uses, including bandages, wound treatment, artificial sutures, body implants, and tissue scaffolding. Nano-emulsion enhances textiles by immobilizing polymeric nanoparticles on their surface by encapsulating active substances [34].

Various antimicrobial textile finishes containing synthetic agents (polyhexamethylene biguanide (PHMB), quaternary ammonium compounds (QACs), metals (metal oxides and salts), triclosan, N-halamines, and peroxy acids) and natural agents (chitosan, alginate, herbal products, essential oils, natural dyes, antimicrobial peptides (AMPs), and sericin) have been commonly used. Depending on the killing or inhibiting mechanism of the antimicrobial drugs, the effects might be long-lasting or short-lived. In addition to the commonly used antibacterial coatings, numerous nanotechnological techniques have been created. Because of their size, shape, surface area, and morphological characteristics, nano-sized antimicrobial agents provide greater benefits and efficacy than conventional antimicrobial agents [35].

Silver-doped polymeric nano-emulsions are widely used in cotton, polyester, and wool as antimicrobial moiety. Silver nanoparticles (AgNPs) are highly poisonous to bacteria but less hazardous to human cells. They are also durable and dyeable. Silver (Ag) nanomaterials are known for their self-cleaning and antibacterial properties [36]. Metal oxide nanoparticles, such as titanium, tin, zinc, gold, and copper, have been used in both natural and synthetic fabrics [37]. Cotton fabric was functionalized with amino-doped titanium dioxide (TiO2) nanoparticles, by employing the two-step sol-gel and hydrothermal process, showed effective antibacterial activity against gram-positive and gram-negative bacteria [38]. Natural chemicals, such as cyclodextrins, can help create sustainable antibacterial textile finishes. The cyclic oligosaccharides have a hydrophilic outer surface and a lipophilic inner chamber for wide applications in textile as core-shell structure [39].

Colloidal solution of silver nanoparticles in polymer shell immobilized on cotton fabric proved to create a significant antibacterial property without affecting the dying properties of fabric [40]. Nano silver has been effectively applied to a variety of natural and synthetic textiles due to its excellent antibacterial properties. Silver nanoparticles integrated into nano shell of polypropylene (PP) significantly enhanced the antibacterial performance compared to microscale polypropylene. The silver has been integrated into the PP by melt mixing in a twin-screw extruder, and fibers have been made. The fibers are found to be reasonable in terms of mechanical qualities, and it has been discovered that more than five times the loading of micron-sized silver particles is necessary to achieve the same antibacterial activity as nano silver against both gram-positive and gram-negative bacteria [41]. The resultant colloidal solution was employed on cotton fabric via pad-dry cure technique using a cross-linker and characterized for further analysis through scanning electron microsocopy (SEM) and transmission electron microsocopy (TEM). The bactericidal action of silver nanoparticles was also studied against bacterial cell wall with significant destruction to cell structure. Nano silver has broad-spectrum antimicrobial properties in a polymeric colloidal solution, thus employed as a nano finishing agent on fabric or fiber surfaces [42, 43].

2.2 Polymeric nano-emulsion in flame-retardant textile finishing

The phrase "Fire retardant fabrics" relates to technical textiles or textile products that cause flame resistance or fire extinguishers' phenomena [44]. Flame-retardant materials are coated on various textile surfaces to protect from flaming by producing char layer on the surface, which protects the textile from burning, as shown in **Figure 3**, respectively. Technical fabrics prioritize performance above esthetics and comfort, which are typically associated with consumer textiles. Flame-retardant materials were coated or immobilized on various textile substrates to provide protection against serious injury in workplace, home, hospital, industry, and military purpose [45].

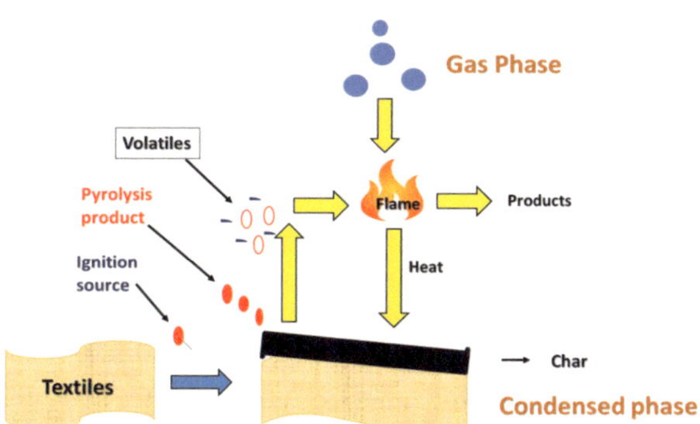

Figure 3.
Nano-emulsion based flame retardant coating in textile.

Flame resistance and flame protection are two different terms. The amount of fire hazard is measured by quantitative characteristics such as time and velocity of burning, and heat release rate upon exposure to flame. Various parameters of textile performance can be evaluated upon exposure to fire as melting and shrinkage of thermoplastic and synthetic fiber textiles, as well as the emission of smoke and poisonous gases during combustion. These parameters are analyzed to consider the thermal protection level of a specific textile [46, 47]. Silicone-based softeners are modified as nano-emulsions with significant flame-retardant properties to polyester fabric by unaltering the mechanical and thermal stability of the fabric [48]. Phosphorus and halogenated components-based polymeric nano formulation can be employed for flame-resistant functionalities to polymeric nanocomposite textile. Similarly, various environment-friendly eco-treatments of textile provide the flame-retardant properties as banana peel and natural essential oil extracts [49, 50].

Lam et al. utilized zinc oxide nanoparticles (ZNO NPs) to impart flame-retardant functional finishes to cotton textile [51]. Similarly, zinc oxide nanoparticles were also used to impart a flame-retardant coating on bleached jute fiber. For this purpose, pad-dry cure method was employed to coat functional finishes by using a cellulose acetate as synthetic binder for durable flame-retardant finishes to jute fiber [52].

Amino-functionalized silica nanoparticles (AFS-NPs) with zinc oxide nanoparticles were immobilized on cotton fabric having an epoxy polymeric shell. The resultant polymeric formulation was characterized through structural, morphological, and particle size analysis. Formulation was dip coated on textile followed by traditional pad-dry cure technique with constant variables as temperature and pressure. Thus, cotton textile showed a synergistic effect of surface hydrophobicity and flame retardancy [53].

2.3 Polymeric nano-emulsion in ultraviolet (UV)-protective textile finishing

Ultraviolet (UV)-protective textile is specifically designed textile to protect humans against risks of harmful UV radiation by sun in the surrounding environment. UV shielding agents and UV-protective finishes are applied to textile fabrics and to impart anti-UV finishing both in woven and nonwoven textiles, as shown in **Figure 4**, respectively [54]. UV radiation, surpassing visible light in energy, can initiate harmful chemical reactions impacting human health. Materials like sportswear, functional textiles, and high-value technical textiles utilize UV protection finishes due to the potential hazards of UV radiation across various spectral bands: ultraviolet-C (UV-C) (190–280 nm) causes skin burns and cancer, ultraviolet-B (UV-B) (280–320 nm) leads to skin damage and eye issues, while ultraviolet A (UV-A) (320–400 nm) results in DNA damage, vision problems, immune system impact, and premature skin aging. Garments and specialized textiles, including wearable sensors and high-altitude clothing, incorporate these finishes to mitigate UV-related risks [55].

The ultraviolet to infrared (IR) electromagnetic (EM) radiation spectrum of sunlight. The UV wavelength is split into three parts: UV-A from 320 to 400 nm, UV-B from 280 to 320 nm, and UV-C from 200 to 280 nm, with the energy per component increasing dramatically. While the ozone layer entirely absorbs the UV-C area, the UV-A and UV-B wavelength regions can reach the earth's surface [56].

Textiles act as a barrier against UV radiation, offering varying levels of protection based on material type, weave, thickness, construction, and chemical finishes [57]. UV-protective clothing relies on its ability to absorb, reflect, and disperse solar

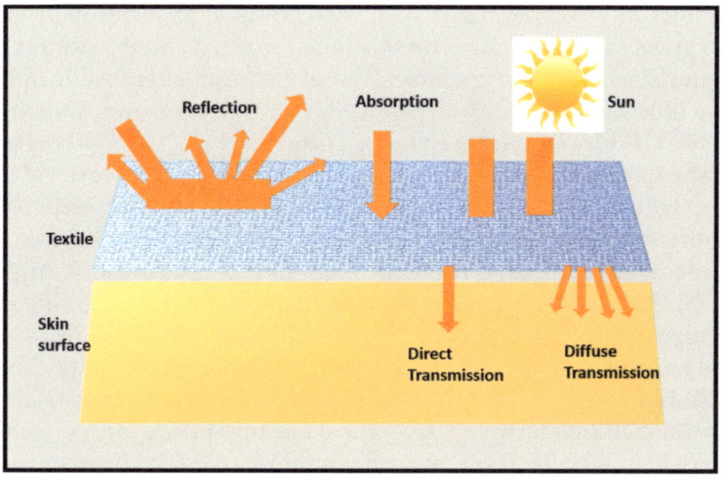

Figure 4.
Nano-emulsion based anti-UV functional textile.

wavelengths, shielding the skin from harmful UV rays. High UV absorption and reflection properties are crucial in these textiles to prevent UV-induced damage [58]. Transmittance, defining the ratio between incident and transmitted UV rays within a specific wavelength range, determines a fabric's UV protection efficacy [59].

When UV radiation interacts with a textile surface, it undergoes various effects. Reflection occurs at the textile surface borders, while absorption takes place as the radiation penetrates the material, transforming into different energy forms. The transmitted portion, known as "transmission," passes through the fabric and reaches the skin [60].

Due to the limitations of traditional textiles in providing protection against diverse environmental threats, the use of nanomaterials in the development of protective textiles has expanded fast [61–65]. Various nanoparticles have been employed to create useful fabrics [66–69]. TiO2 nanomaterial for UV protection textiles is one of the specialized uses of nanomaterials for protective textiles [70]. UV protection fabrics are required in these high ultraviolet radiation (UVR) hazard-prone industrial work environments [71]. UV protection fabrics can also be employed in other industrial sectors such as aircraft and automobiles [72]. UV protection textiles may be made utilizing a variety of methods, including fiber/yarn selection, weaving parameters, coating and textile finishing techniques, and so on [73]. Natural extracts of fruits, vegetables, flowers, etc., have inherent anti-UV properties that are employed on textile surface by encapsulating them in a polymer shell to impart multifunctional characteristics' finishing to the cotton, polyester, and wool [74].

Inorganic nanoparticles, such as TiO2, ZnO, silicon dioxide (SiO2), copper oxide (CuO), aluminum oxide (Al_2O_3)), and reduced graphene oxide (rGO), are considered efficient materials for developing thermal and chemical stability, anti-UV, and nontoxicity [75, 76]. Metal oxide nanostructures absorb UV photons equivalent to or greater than their band gap energy via modifying textiles and polymers. The band gap of nanostructures is in the UV region of the solar spectrum, and it has a high potential to absorb UV radiation due to its large specific surface area [77, 78]. Thus, UV radiation absorption may be increased depending upon surface area and size at nanoscale [79]. Nano-sized TiO2 exhibits superior UV-protection properties compared to micron-sized pigment TiO2, with stronger absorption properties. Its high absorption capacity and refractive index make it suitable for textiles and polymers. Nano-sized TiO2 can reduce UV-A absorption near visible light due to its negative blue shift on the absorption edge [80]. Emam and Abdelhameed investigated the UV-blocking properties of cotton and silk textiles coated with Materials of the Lavoisier Institute (MIL) metal organic framework (MOF). The UV protection factor (UPF) of treated textiles increased linearly with MIL-MOF and metal content. The incorporation of 10.4 g of MIL-MOF/kg on the textile surface was adequate to obtain outstanding UV-blocking capability. Approximately 40% of the MIL-MOFs were washed away during the first five washing cycles; yet, cleaned materials demonstrated effective UV blocking (UPF 26.7–36.2). *In situ* synthesis was used to attach MOFs to the surface of textiles [81]. Silva et al. discovered UV-blocking properties of chitosan-coated gold nanoparticles (AuNPs) on soybean protein-based knitted rib textiles. Because of the low concentration gold nanoparticle immobilization on soybean fabric, pure soybean knitted fabric has a UPF of approximately 7, whereas gold nanoparticle incorporating fabric has a UPF of around 14 [82]. Chitosan-treated cloth had a UPF of approximately 62, but gold nanoparticle-containing chitosan coating had a UPF of around 289 (almost 40-fold higher than pure fabric) with little UV-A and UV-B transmission. Even after numerous washing cycles, gold nanoparticle-containing chitosan-coated cloth has exceptional UV-blocking properties [83].

Nanomaterials utilized in the creation of UV protection textiles can be synthesized using a variety of techniques, such as the sol-gel method, chemical vapor deposition (CVD), and emulsion method [84]. Ran et al. developed UV-protected cotton textiles using a CuO/BiVO4 (BVO) (bismuth vanadate) nanocomposite photo catalyst immobilized by polydopamine (PDA) templating [85].

The utilization of ethyl cellulose (EC) nanoparticles as nanocarriers of active or lipid-soluble chemicals, followed by their deposition on cotton textiles, developed a UV-protective textile. Thermal behavior assessment, estimate of work of adhesion (WA), and wash resistance tests were also used to investigate the EC/cotton affinity and attachment mechanism of EC nanoparticles on cotton substrate. It is hypothesized that entanglement of polymeric chains during EC nanoparticle deposition on cotton fabric improves EC nanoparticle adherence on cotton substrate. The functioning of cotton textile was evaluated using the UV protection factor (UPF) measurements that revealed a high UPF value (UPF = 45) [86].

2.4 Polymeric nano-emulsion in water-repellent and hydrophobic textile finishing

Water-resistant textiles are frequently high-density woven fabrics consisting of very fine threads or conventional materials treated with a hydrophobic finishing, as illustrated in **Figure 5** [87]. Waterproofing is the characteristic of a substance that prevents fluids from penetrating it. Waterproofness may be assessed using one of the two methods: one that simulates rain and the other (more frequent) that submits the cloth to hydrostatic pressure. Specific surface finishing treatments are used to create water- resistant fabrics. Waterproof materials are often achieved by surface finishing procedures such as coating [88]. Covering is a broad phrase that refers to the application of one or more layers of adhering polymeric goods to one or both sides of a textile material to produce a film. Technologies employed for polymeric nano-emulsion in hydrophobic functional finishing of textiles are as follows [89, 90].

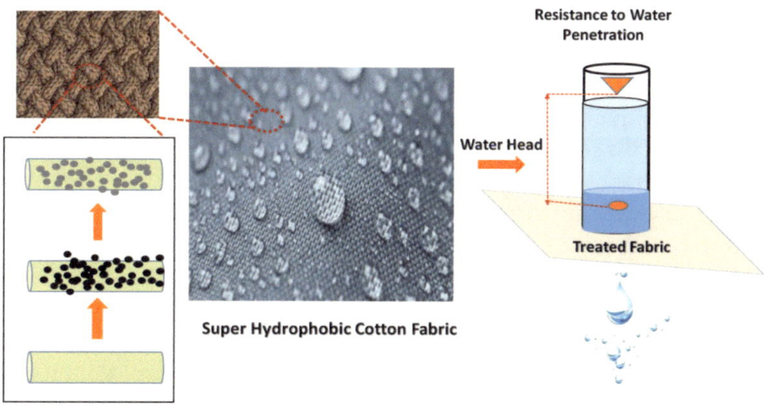

Figure 5.
Nano-emulsion based hydrophobic finishing in textile.

- Coating technology utilizes direct layering or superficial impregnation of polymers, commonly applied as either a paste or a high-viscosity liquid during the final phase of waterproof material production, yielding ultrathin coatings spanning 10–100 μm.

- Laminating technology initiates with the creation of a primary laminating layer (membrane or foam) subsequently distributed onto textile surfaces. With membranes, like polytetrafluoroethylene (PTFE) at approximately 10-μm thinness, ensuring the final film thickness remains within the 10–100 μm range, this technique provides waterproofing to the fabric through a layered application process [91].

Waterproof and water-repellent textiles find extensive functional applications across diverse sectors like agriculture, biomedical, food and packaging industry [92]. Waterproof textile serves as breathable ground covers, tents, root protective bags, greenhouse covers, and tree shelters in agriculture. Additionally, these textiles are utilized in agricultural structures for leakproof water and liquid fertilizer tanks, flexible water tanks, and packaging products [93]. Architectural textiles, characterized by their lightweight and flexibility, durability and coatings or impregnation, are used to enhance energy efficiency and cost-effectiveness while fostering innovative architectural design [94]. Textile membranes, composed of high-strength woven fabrics like glass fibers, polyester (PES), or polyethylene coated with polyvinyl chloride (PVC), silicone, or PTFE, facilitate temporary applications, such as tents, clear-span structures, tension fabric structures, and air structures, serving purposes ranging from exhibitions and leisure activities to short-term commercial spaces and storage facilities [95].

Hydrophobic chemicals, such as paraffin waxes, silicones, silanes, and fluorinated polymers, are traditionally utilized to impart water-resistant functionalities [96]. However, current studies indicate the feasibility of employing nano substances such as SiO2, ZnO, and TiO2 [97]. A standard parameter for a superhydrophobic water-repellent coating on fabric is a contact angle of a drop greater than 150°, which is known as the "Lotus-Effect" [98]. Hydrophobic/superhydrophobic surfaces on textile are developed by producing rough structures on a surface or by changing a rough surface with low surface free energy compounds. When water-repellent nano finishes are applied to a textile, the surface becomes hydrophobic and drops of water readily fall off [99]. The water-repellent nano coatings greatly retain the fabric's air permeability and breathability [100].

The sol-gel technique produced a silane linkage by using silica as precursors. These sol-gel coatings contain silica nanoparticles (200 nm), a perfluorooctanoic acid-free (PFOA) fluoropolymer, 3-glycidoxypropyltrimethoxysilane, and tetraethyl orthosilicate. The dip-dry care approach was used to apply perfluorooctanoic acid-free fluoropolymer-coated silica nanoparticles to the cotton fabric surface. The substrate exhibited significant contact angles of 154° and 120° for water and n-dodecane, indicating strong liquid repellency [101]. The layer-by-layer approach, with electrostatic self-assembly, was used to coat a cotton fabric, resulting in the creation of poly (diallyldimethylammonium chloride) (PDDA) and ZnO/SiO_2 colloidal nanocomposite solutions. The number of deposition layers on the fibers showed the presence of a ZnO/SiO_2 nanocomposite [102]. The titanium (IV) butoxide nanoparticles were produced with boric acid as the functional ingredient. The pad-dry process facilitates the application of sols to cotton textiles. This process improved the water-resistant properties of the resulting cotton fabric. Furthermore, the treated fabric demonstrated flame-retardant properties and high thermal stability [103]. Spraying an aqueous dispersion over the silk fabric provides the appropriate water- and oil-resistant characteristics. Further dispersion of silica nanoparticles, quaternary ammonium silane salt, organic fluoropolymer, and alkoxy silanes created the highest contact angle on the textile surface due to immobilization on textile surface [104].

2.5 Polymeric nano-emulsion in conductive textile finishing

Conductive textiles are electrically conductive for various industrial applications. Such textiles may be created by employing conductive fibers or putting conductive layers over nonconductive fabrics [105]. The phrase "electrically conductive textiles" refers to a diverse spectrum of textile fiber-based products with considerably varying specific electrical conductivity. Electrically conductive textiles include conductive fibers, yarns, fabrics, and finished goods [106]. A conductive thread that is commercially accessible is often either a solid metal wire, such as copper or stainless steel, or a nonconductive thread covered with a B1 mm thick metal film, usually silver [107]. Cotton, polyester, and nylon threads were rendered electrically conductive by covering their surfaces with silver nanowire meshes. The weight of the nanowire coating is less than 5% that of normal silver thin-film coatings used in commercial conductive thread [108].

Electrically conductive textile has attracted a large commercial market in wearable electronic devices, such as pressure and gas sensors, energy storage devices, wearable warm textile, and electromagnetic interference (EMI) shielding [109]. Layer-by-layer self-assembling of electrically charged ions via dip-coating techniques is employed to develop various types of conductive functional textile substrates, such as cotton, polyester, and wool [110].

Various types of conductive materials as metal nanoparticles, MXenes, reduced graphene oxide, and carbon nanotubes (CNTs) were employed on textiles via dip-coating, layer-by-layer technique, and inkjet printing for conductive smart textile applications [111]. Conductive textiles can be made directly from polymer composite fibers that contain conductive agents [112–114]. Seyedin et al., for example, created highly conductive polyurethane fibers by combining polyurethane with poly(3,4-ethylenedioxythiophene), poly(styrenesulfonate) (PEDOT:PSS) in dimethyl sulfoxide (DMSO), followed by wet-spinning for textiles in the use of strain sensors [114]. Graphene has significant applications in developing conductive textile by coating two-dimensional (2D) thin-film structure that provides excellent mechanical and thermal stability as well as durability without altering the inherent textile properties [115].

Conducting fibers plays a key role in electronic textile, due to their unique properties such as lightweight, flexibility, and durability, depending on the area of application [116]. Such fibers can be employed as electrode material, electronic transistors, light-emitting diodes (LEDs), and solar cells. Conductive textile fibers are being employed in the aviation industry as a stronger and more flexible weight-saving material and replace the heavy metal conductors by lightweight functional fabric, which reduces the cost and fuel consumption [117]. Electrically conductive fabrics have been investigated for use in a variety of applications due to their favorable features in terms of ESD (electrostatic discharge) coating, EMI protection, radio frequency protection, and thermal expansion [118]. Various methods for producing electrically conductive textiles, including conductive fiber/yarn manufacturing, coating, and binding of conductive fiber before and after woven textile manufacturing and weaving of textile, have been developed. Thus, conductive textile has a variety of industrial applications including antistatic, EMI shielding, electrical applications, and infrared absorption [106].

Silver, copper, and zinc nanoparticles have been extensively studied by researchers due to their unique antibacterial, conductivity, and UPF properties [119]. Adding nanoparticles of AgO, ZnO, and Cu to cotton fabric enhances its qualities, including antimicrobial, UPF, and conductivity [120]. A cost-effective and efficient method for modifying cotton textiles has been developed, resulting in faster operations. Cotton textiles loaded with AgNPs, AgNPs/ZnONPs, or AgNPs/ZnONPs/CuNPs formed metallic and metal oxide nanoparticles. The presence of trimetallic nanoparticles inside microstructure of cotton provides long-lasting antibacterial, UV protection, and conductivity capabilities, resulting in multifunctional textiles. Nanoparticles were produced and stabilized independently using polymethyl compound (PMC) and functionalized polyethyleneimine (FPEI) in a single bath. Addition of PMC or FPEI compounds to metal compound solutions changes their color from colorless to yellow or dark green. Ultraviolet-visible (UV-vis) spectra show a maximum surface plasmon peak at 410–415, indicating effective synthesis of AgNPs stabilized by PMC or FPEI chains. Nanoparticles are effectively deposited on the surface of cellulose textiles, resulting in changes in crystalline structure. Fabrics treated according to design demonstrate multifunctional qualities and performance. After 20 washing cycles, the treated cotton fabric retains good antibacterial qualities, as well as great ultraviolet and electrical conductivity [121].

2.6 Polymeric nano-emulsion in moisture-wicking textile finishing

Wicking is the process of absorbing a liquid by capillary forces in a textile structure. Its behavior is determined by the fibers' wettability, the likelihood of film formation, and the existence of capillary gaps [122]. In high-performance or critical conditions, liquid management in textiles is vital. Wicking in textiles is a two-phase imbibition phenomenon, in which a liquid (for example, water) displaces a gas or air in a textile structure. It happens when the fibers are wettable, and water may move through the gaps between the fibers via capillary action [123]. Wicking is defined as the nonsaturated flow of liquid through the textile, i.e., the two-phase flow of liquid and air. This liquid movement can occur concurrently with vapor transport. Wicking can occur with a restricted reservoir of liquid (for example, liquid from a droplet) or an infinite reservoir of liquid (for example, imbibition) [124].

Moisture-wicking technology in textiles allows for rapid drying in hot and humid environments by transferring liquid water and vapor from the body to the environment, creating a pleasant microclimate for the wearer [125, 126]. Researchers are continuously increasing the moisture-wicking qualities of fabrics to meet the increased need for rapid drying performance [127]. Commercially available fabrics provide varying amounts of personal moisture wicking. The most prevalent moisture-wicking technology is high wicking fabrics made of profiled fibers, which use greater capillary force to quickly remove perspiration from the skin [128].

Application of resin is considered a conventional treatment to impart wrinkle-free or crease-resistant properties to textiles, as resin treatment affects the textile breathability, mechanical stability, textile dyeing properties, and washing fastness [129]. Some researchers have used nanostructured materials with wrinkle-resistant properties to address the restrictions of employing resins. Polymeric nano formulation employed as crease-resistant finishing via cross-linking agents stimulates the durability as well as improves the textile inherent properties [130].

Synthetic reagents such as sodium hypophosphite and titania oxide nanoparticles were also employed on textile to improve the wrinkle-free treatment as functional finishing agents. Titania oxide nanoparticles were also employed as co-catalysts to enhance the functional properties and reduce the adverse effects. Furthermore, nano-$TiO2$ is extremely resistant to high temperatures and pressures. Because of its enormous surface area, nano-$TiO2$ has enough active sites to achieve acceptable reaction rates [131].

Lu et al. used ion-gel technology to create hydroxypropyl chitosan nanoparticles, applied on silk fabric in the presence of 1,2,3,4-butanetetracarboxylic acid (BTCA) and sodium hypophosphite using conventional padding followed by drying and curing at certain temperature. According to Lu et al., silk fabric with chitosan nanoparticles improved its wrinkle resistance [132]. In addition, Zulfiqar et al. synthesized chitosan nanoparticles using ionic gelation process and immobilized the polymeric nanoparticles to cotton fabric using the pad-dry-cure method. Thus, increased crease resistance and anti-wrinkle property are due to the penetration of polycationic nanoparticles into the interphase textile fabric [133].

3. Conclusion

In conclusion, the evolution of functional textiles through the application of nanotechnology has revolutionized the textile industry, addressing challenges related to hygiene, safety, and performance. This chapter comprehends the significant role of polymeric nano-emulsions in functional textile finishes, providing a controlled and durable application without compromising the inherent characteristics of the original textiles. The diversified applications discussed, ranging from antimicrobial, flame-retardant, UV-protective, and waterproof textiles to conductive and moisture-wicking textiles, showcase the versatility and transformative impact of advanced textile finishing methods. These innovations not only meet contemporary challenges posed by bacterial infections, environmental factors, and specialized functional requirements but also contribute to the dynamic landscape of textile technology. With the appearance of new technologies that may add specific functionalities and noticeable characteristics to materials, the textile industry is experiencing a new revolution. The integration of nanotechnology emerges as a key driver in the continuous improvement of textile functionalities, opening new frontiers for exploration and application across various industries.

Acknowledgements

Sana Javaid acknowledges the support of UW Research Directorate.

Conflict of interest

The authors declare no conflict of interest.

Author details

Sana Javaid[1*], Nadia Saleem[1] and Shafi Ur Rehman[2]

1 Department of Chemistry, University of Wah, Wah Cantt, Pakistan

2 School of Chemical and Materials Engineering (SCME), National University of Sciences and Technology (NUST), Islamabad, Pakistan

*Address all correspondence to: sana.javaid@uow.edu.pk

References

[1] Chaleshtari Z, Zhou M, Foudazi RJJOAP. Nanoemulsion polymerization and templating: Potentials and perspectives. Journal of Applied Physics. 2022;**131**(15):150902

[2] Mason TGJM. Emulsiogenesis and the emergence of Nanoemulsions. Matter. 2019;**1**(3):542-546

[3] Calvo P, Vila-Jato JL, Alonso MJJJOPS. Comparative in vitro evaluation of several colloidal systems, nanoparticles, nanocapsules, and nanoemulsions, as ocular drug carriers. Journal of Pharmaceutical Sciences. 1996;**85**(5):530-536

[4] McClements DJ, Jafari SM. General aspects of nanoemulsions and their formulation. In: Nanoemulsions. United States: Academic Press; 2018. pp. 3-20

[5] Araújo F et al. Development and characterization of parenteral nanoemulsions containing thalidomide. European Journal of Pharmaceutical Sciences. 2011;**42**(3):238-245

[6] Vatsraj S, Chauhan K, Pathak HJJON. Formulation of a novel nanoemulsion system for enhanced solubility of a sparingly water soluble antibiotic, clarithromycin. Journal of Nanoscience. 2014;**2014**:268293

[7] Shakeel F, M.S.J.P.d. Faisal, and technology, Nanoemulsion: A promising tool for solubility and dissolution enhancement of celecoxib. Pharmaceutical Development and Technology. 2010;**15**(1):53-56

[8] Sonneville-Aubrun O et al. Nanoemulsions: A new vehicle for skincare products. 2004;**108**:145-149

[9] Terescenco D et al. From interface towards organised network: Questioning the role of the droplets arrangements in macroscopically stable O/W emulsions composed of a conventional non-ionic surfactant, TiO2 particles, or their mixture. 2019;**578**:123630

[10] Fryd MM, Mason TGJAROPC. Advanced nanoemulsions. Annual Review of Physical Chemistry. United States: Academic Press; 2012;**63**:493-518

[11] Håkansson A, Rayner M. General principles of nanoemulsion formation by high-energy mechanical methods. In: Nanoemulsions. Elsevier; 2018. pp. 103-139

[12] Solans C et al. Nano-emulsions. Current Opinion in Colloid & Interface Science. 2005;**10**(3-4):102-110

[13] Solans C, Solé I. Nano-emulsions: Formation by low-energy methods. Current Opinion in Colloid & Interface Science. 2012;**17**(5):246-254

[14] Joshi M, Bhattacharyya A. Nanotechnology–a new route to high-performance functional textiles. Textile Progress. 2011;**43**(3):155-233

[15] Maity S, Singha K, Pandit P. Functional and Technical Textiles. India: Woodhead Publishing; 2023

[16] Bashari A et al. Functional finishing of textiles via nanomaterials. Nanomaterials in the Wet Processing of Textiles. 2018:1-70

[17] Flavel BS et al. Patterned attachment of carbon nanotubes to silane modified silicon. Carbon. 2007;**45**(13):2551-2558

[18] Dastjerdi R, Montazer M. A review on the application of inorganic nano-structured materials in the modification of textiles: Focus on anti-microbial properties. Colloids and Surfaces B: Biointerfaces. 2010;**79**(1):5-18

[19] Riaz S et al. Functional finishing and coloration of textiles with nanomaterials. Coloration Technology. 2018;**134**(5):327-346

[20] Abou Elmaaty TM et al. Recent advances in functionalization of cotton fabrics with nanotechnology. Polymers. 2022;**14**(20):4273

[21] Shah MA et al. Applications of nanotechnology in smart textile industry: A critical review. Journal of Advanced Research. 2022;**38**:55-75

[22] Antunes JC et al. Bioactivity of chitosan-based particles loaded with plant-derived extracts for biomedical applications: Emphasis on antimicrobial Fiber-based systems. Marine Drugs. 2021;**19**(7):359. DOI: 10.3390/md19070359

[23] Ferrero F, Periolatto M. Antimicrobial finish of textiles by chitosan UV-curing. Journal of Nanoscience and Nanotechnology. 2012;**12**(6):4803-4810

[24] Deng Z et al. Applications of chitosan-based biomaterials: A focus on dependent antimicrobial properties. Marine Life Science & Technology. 2020;**2**(4):398-413

[25] Gulati R, Sharma S, Sharma RK. Antimicrobial textile: Recent developments and functional perspective. Polymer Bulletin. 2022;**79**(8):5747-5771

[26] Iyigundogdu ZU et al. Developing novel antimicrobial and antiviral textile products. Applied Biochemistry and Biotechnology. 2017;**181**:1155-1166

[27] Javaid S et al. Cefotaxime loaded polycaprolactone based polymeric nanoparticles with antifouling properties for in-vitro drug release applications. Polymers. 2021;**13**(13):2180

[28] Nadi A et al. Evolution in the surface modification of textiles: A review. Textile Progress. 2018;**50**(2):67-108

[29] Perelshtein I, Perkas N, Gedanken A. Ultrasonic coating of textiles by antibacterial and antibiofilm nanoparticles. In: Handbook of Ultrasonics and Sonochemistry. Singapore: Springer Science Business Media; 2016. pp. 967-993

[30] Nawab R et al. Review featuring the use of inorganic nano-structured material for anti-microbial properties in textile. Polymer Bulletin. 2023;**80**(7):7221-7245

[31] Andra S et al. Emerging nanomaterials for antibacterial textile fabrication. Naunyn-Schmiedeberg's Archives of Pharmacology. 2021;**394**:1355-1382

[32] Diedrich C et al. Optimized chitosan-based Nanoemulsion improves luteolin release. Pharmaceutics. 2023;**15**(6):1592. DOI: 10.3390/pharmaceutics15061592

[33] Massella D et al. Bio-functional textiles: Combining pharmaceutical nanocarriers with fibrous materials for innovative dermatological therapies. Pharmaceutics. 2019;**11**(8):403

[34] Javaid S et al. Layer-by-layer self-assembled dip coating for antifouling functionalized finishing of cotton textile. Polymers. 2022;**14**(13):2540

[35] Arik B. Common and nano-antimicrobial textile finishes. In:

Antimicrobial Textiles from Natural Resources. Cambridge: Woodhead Publishing; 2021. pp. 87-117

[36] Butola B, Mohammad F. Silver nanomaterials as future colorants and potential antimicrobial agents for natural and synthetic textile materials. RSC Advances. 2016;**6**(50):44232-44247

[37] Román LE et al. Antibacterial cotton fabric functionalized with copper oxide nanoparticles. Molecules. 2020;**25**(24):5802

[38] Zhang G et al. Study on the photocatalytic and antibacterial properties of TiO2 nanoparticles-coated cotton fabrics. Materials. 2019;**12**(12):2010

[39] Singh N, Sahu O. Sustainable cyclodextrin in textile applications. In: The Impact and Prospects of Green Chemistry for Textile Technology. Cambridge: Woodhead Publishing; 2019. pp. 83-105

[40] Perera S et al. Morphological, antimicrobial, durability, and physical properties of untreated and treated textiles using silver-nanoparticles. Colloids and Surfaces A: Physicochemical and Engineering Aspects. 2013;**436**:975-989

[41] Jeong SH, Yeo SY, Yi SC. The effect of filler particle size on the antibacterial properties of compounded polymer/silver fibers. Journal of Materials Science. 2005;**40**:5407-5411

[42] Jafari N et al. Effect of silver particle size on color and antibacterial properties of silk and cotton fabrics. Fibers and Polymers. 2016;**17**:888-895

[43] Pereira C et al. Nanoengineered textiles: From advanced functional nanomaterials to groundbreaking high-performance clothing. In: Handbook of Functionalized Nanomaterials for Industrial Applications. Cambridge: Woodhead Publishing; 2020. pp. 611-714

[44] Gaan S et al. Flame retardant functional textiles. In: Functional Textiles for Improved Performance, Protection and Health. Cambridge: Woodhead Publishing; 2011. pp. 98-130

[45] Alongi J, Horrocks AR, Carosio F, Malucelli G, editors. Regulatory and testing requirements for flame retardant textile applications. Update on flame retardant textiles: State of the art, environmental issues and innovative solutions. Shawbury, UK: University of Bolton Institutional Repository (UBIR); 2013. pp. 54-122

[46] Carosio F et al. Flame retardancy of polyester and polyester–cotton blends treated with caseins. Industrial & Engineering Chemistry Research. 2014;**53**(10):3917-3923

[47] Horrocks AR. Flame retardant challenges for textiles and fibres: New chemistry versus innovatory solutions. Polymer Degradation and Stability. 2011;**96**(3):377-392

[48] Parvinzadeh Gashti M, Yousefpour Navid M, Hossein Rahimi M. Effects of coating of nano- and microemulsion silicones on thermal properties and flammability of polyethylene terephthalate textile. Pigment & Resin Technology. 2013;**42**(1):34-44

[49] Khandual A. Green flame retardants for textiles. In: Muthu SS, Gardetti MA, editors. Green Fashion. Vol. 2. Singapore: Springer Singapore; 2016. pp. 171-227

[50] Malucelli G et al. Biomacromolecules as novel green flame retardant systems

for textiles: An overview. RSC Advances. 2014;**4**(86):46024-46039

[51] Lam Y, Kan CW, Yuen C. Flame-retardant finishing in cotton fabrics using zinc oxide co-catalyst. Journal of Applied Polymer Science. 2011;**121**(1):612-621

[52] Samanta AK et al. Fire retardant finish of jute fabric with nano zinc oxide. Cellulose. 2017;**24**:1143-1157

[53] Mohamed AL, El-Sheikh MA, Waly AI. Enhancement of flame retardancy and water repellency properties of cotton fabrics using silanol based nano composites. Carbohydrate Polymers. 2014;**102**:727-737

[54] Gambichler T. Ultraviolet protection of clothing. In: Functional Textiles for Improved Performance, Protection and Health. Cambridge: Woodhead Publishing; 2011. pp. 45-63

[55] Chakraborty J. Enhancing UV protection of cotton through application of novel UV absorbers. Journal of Textile and Apparel, Technology and Management. 2014;**9**(1):1-13

[56] Chen-Yu JH, Wong S-W. Effects of repeated treatments of perspiration, weathering, and laundering on the ultraviolet protection of a naturally colored lightweight cotton fabric. Clothing and Textiles Research Journal. 2017;**35**(2):128-140

[57] Attia NF et al. Influence of textile fabrics structures on thermal, UV shielding, and mechanical properties of textile fabrics coated with sustainable coating. Journal of Natural Fibers. 2021;**18**(12):2189-2196

[58] Attia NF et al. Recent advances of sustainable textile fabric coatings for UV protection properties. Coatings. 2022;**12**:1597. DOI: 10.3390/coatings12101597

[59] Gies H et al. UV protection by clothing: An intercomparison of measurements and methods. Health Physics. 1997;**73**(3):456-464

[60] Scott RA. Textiles for Protection. Cambridge: Woodhead Publishing; 2005

[61] Irzmańska E et al. Evaluation of the mechanical parameters of ultrasonically welded textile composite structures for protective footwear. Fibres & Textiles in Eastern Europe. 2019;**27**, 3(135):99-105

[62] Lee DT et al. Water-stable chemical-protective textiles via euhedral surface-oriented 2D Cu–TCPP metal-organic frameworks. Small. 2019;**15**(10):1805133

[63] Rather LJ et al. Coloration, UV protective, and antioxidant finishing of wool fabric via natural dye extracts: Cleaner production of bioactive textiles. Environmental Progress & Sustainable Energy. 2019;**38**(5):13187

[64] Zhang L et al. A self-protective, reproducible textile sensor with high performance towards human–machine interactions. Journal of Materials Chemistry A. 2019;**7**(46):26631-26640

[65] Rivero PJ et al. Nanomaterials for functional textiles and fibers. Nanoscale Research Letters. 2015;**10**:1-22

[66] Qiu Q et al. Functional nanofibers embedded into textiles for durable antibacterial properties. Chemical Engineering Journal. 2020;**384**:123241

[67] Zhao Z et al. Cellulose micro-dissolution by N-methylmorpholine N-oxide as a facile route for magnetic functional cotton textiles. Cellulose. 2020;**27**:1817-1828

[68] Rodrigues AG et al. Functional textiles impregnated with biogenic silver nanoparticles from Bionectria ochroleuca and its antimicrobial activity. Biomedical Microdevices. 2019;**21**:1-10

[69] Lu X et al. A multi-functional textile that combines self-cleaning, water-proofing and VO2-based temperature-responsive thermoregulating. Solar Energy Materials and Solar Cells. 2017;**159**:102-111

[70] Ryu S-Y, Park M-K, Kwak S-Y. Silver-titania/polyurethane composite nanofibre mat for chemical and biological warfare protection. International Journal of Nanotechnology. 2013;**10**(8/9):771-788

[71] Bauer S. Photoprotection by workwear: Ultraviolet protection factors for artificial radiation from welding arcs. Photochemistry and Photobiology. 2020;**96**(4):917-925

[72] Sezer Hicyilmaz A, Altin Y, Bedeloglu A. Polyimide-coated fabrics with multifunctional properties: Flame retardant, UV protective, and water proof. Journal of Applied Polymer Science. 2019;**136**(23):47616

[73] Mondal S. Nanomaterials for UV protective textiles. Journal of Industrial Textiles. 2022;**51**(4_suppl):5592S-5621S

[74] Joshi S et al. Effect of onion (Allium cepa L.) Peel extract-based Nanoemulsion on anti-microbial and UPF properties of cotton and cotton blended fabrics. Journal of Natural Fibers. 2022;**19**(14):8345-8354

[75] Selishchev D et al. Effect of preparation method of functionalized textile materials on their photocatalytic activity and stability under UV irradiation. Chemical Engineering Journal. 2013;**224**:114-120

[76] Hasani M, Montazer M. Electro-conductivity, bioactivity and UV protection of graphene oxide-treated cellulosic/polyamide fabric using inorganic and organic reducing agents. The Journal of The Textile Institute. 2017;**108**(10):1777-1786

[77] Kanakaraju D, Chandrasekaran A. Recent advances in TiO2/ZnS-based binary and ternary photocatalysts for the degradation of organic pollutants. Science of the Total Environment. 2023;**868**:161525

[78] Sang Y, Liu H, Umar A. Photocatalysis from UV/Vis to near-infrared light: Towards full solar-light spectrum activity. ChemCatChem. 2015;**7**(4):559-573

[79] Montazer M, Amiri MM, Malek RMA. In situ synthesis and characterization of nano ZnO on wool: Influence of nano photo reactor on wool properties. Photochemistry and Photobiology. 2013;**89**(5):1057-1063

[80] Farouk A et al. Sol-gel-derived inorganic-organic hybrid polymers filled with zno nanoparticles as an ultraviolet protection finish for textiles. Autex Research Journal. 2010;**10**(3):58-63

[81] Emam HE, Abdelhameed RM. Anti-UV radiation textiles designed by embracing with nano-MIL (Ti, In)–metal organic framework. ACS Applied Materials & Interfaces. 2017;**9**(33):28034-28045

[82] Borase HP et al. Phytolatex synthesized gold nanoparticles as novel agent to enhance sun protection factor of commercial sunscreens. International Journal of Cosmetic Science. 2014;**36**(6):571-578

[83] Silva IO et al. Multifunctional chitosan/gold nanoparticles coatings for biomedical textiles. Nanomaterials. 2019;**9**(8):1064

[84] Meirowitz RE. 8 – Coating processes and techniques for smart textiles. In: Hu J, editor. Active Coatings for Smart Textiles. Cambridge: Woodhead Publishing; 2016. pp. 159-177

[85] Ran J et al. Immobilizing CuO/BiVO4 nanocomposite on PDA-templated cotton fabric for visible light photocatalysis, antimicrobial activity and UV protection. Applied Surface Science. 2019;**493**:1167-1176

[86] Vílchez-Maldonado S et al. UV protective textiles by the deposition of functional ethylcellulose nanoparticles. Cellulose. 2014;**21**:2133-2145

[87] Liu M et al. Hydrophobic, breathable cellulose nonwoven fabrics for disposable hygiene applications. Carbohydrate Polymers. 2022;**288**:119367

[88] Wei DW et al. Superhydrophobic modification of cellulose and cotton textiles: Methodologies and applications. Journal of Bioresources and Bioproducts. 2020;**5**(1):1-15

[89] Xue C-H et al. Superhydrophobic surfaces on cotton textiles by complex coating of silica nanoparticles and hydrophobization. Thin Solid Films. 2009;**517**(16):4593-4598

[90] Bahners T et al. Recent approaches to highly hydrophobic textile surfaces. Journal of Adhesion Science and Technology. 2008;**22**(3-4):285-309

[91] Loghin C et al. Introduction to waterproof and water repellent textiles. In: Waterproof and Water Repellent Textiles and Clothing. Cambridge: Woodhead Publishing; 2018. pp. 3-24

[92] Sharma N et al. A substantial role of Agro-textiles in agricultural applications. Frontiers in Plant Science. 2022;**13**:895740

[93] Yaseen M et al. Use of textile waste water along with liquid NPK fertilizer for production of wheat on saline sodic soils. International Journal of Phytoremediation. 2016;**18**(5):502-508

[94] Mullangi D et al. Super-hydrophobic covalent organic frameworks for chemical resistant coatings and hydrophobic paper and textile composites. Journal of Materials Chemistry A. 2017;**5**(18):8376-8384

[95] Houtman R. Materials used for architectural fabric structures. In: Fabric Structures in Architecture. Cambridge: Woodhead Publishing; 2015. pp. 101-121

[96] Espanhol-Soares M et al. Super-hydrophobic coatings on cotton fabrics using sol–gel technique by spray. Journal of Sol-Gel Science and Technology. 2020;**95**(1):22-33

[97] Wu M, Wang M, Ge M. Investigation into the performance and mechanism of SiO2 nanoparticles and starch composite films. The Journal of The Textile Institute. 2009;**100**(3):254-259

[98] Spaeth M, Barthlott W. Lotus-effect®: Biomimetic super-hydrophobic surfaces and their application. Advances in Science and Technology. 2009;**60**:38-46

[99] Ramaratnam K et al. Ultrahydrophobic textiles using nanoparticles: Lotus approach. Journal of Engineered Fibers and Fabrics. 2008;**3**(4):155892500800300402

[100] Vasiljević J et al. Fabrication of the hierarchically roughened bumpy-surface topography for the long-lasting highly oleophobic "lotus effect" on cotton fibres. Cellulose. 2016;**23**(5):3301-3318

[101] Ren H-T et al. Superhydrophobic TiN-coated cotton fabrics with nanoscale

roughness and photothermal self-healing properties for effective oil–water separation. ACS Applied Nano Materials. 2023;**6**(13):11925-11933

[102] Abd El-Hady M, Sharaf S, Farouk A. Highly hydrophobic and UV protective properties of cotton fabric using layer by layer self-assembly technique. Cellulose. 2020;**27**:1099-1110

[103] Bentis A, Boukhriss A, Gmouh S. Flame-retardant and water-repellent coating on cotton fabric by titania–boron sol–gel method. Journal of Sol-Gel Science and Technology. 2020;**94**:719-730

[104] Hu J et al. Durable and super-hydrophilic/underwater super-oleophobic two-dimensional MXene composite lamellar membrane with photocatalytic self-cleaning property for efficient oil/water separation in harsh environments. Journal of Membrane Science. 2021;**637**:119627

[105] Temesgen AG, Fırat TO. Conductive fabrics and their application in technical textiles via Arduino uno microcontroller. Abyssinia Journal of Engineering and Computing. 2022;**2**(2):51-59

[106] Grancarić AM et al. Conductive polymers for smart textile applications. Journal of Industrial Textiles. 2018;**48**(3):612-642

[107] Atwa Y, Maheshwari N, Goldthorpe IA. Silver nanowire coated threads for electrically conductive textiles. Journal of Materials Chemistry C. 2015;**3**(16):3908-3912

[108] Maheshwari N. Silver Nanowire Coatings for Electrically Conductive Textiles. Cambridge: Woodhead Publishing; 2016

[109] Matsuhisa N et al. Printable elastic conductors with a high conductivity for electronic textile applications. Nature Communications. 2015;**6**(1):7461

[110] Chen X et al. Flame-retardant, electrically conductive and antimicrobial multifunctional coating on cotton fabric via layer-by-layer assembly technique. RSC Advances. 2016;**6**(33):27669-27676

[111] Zhu S et al. Multi-functional and highly conductive textiles with ultra-high durability through 'green' fabrication process. Chemical Engineering Journal. 2021;**406**:127140

[112] Lee S et al. Ag nanowire reinforced highly stretchable conductive fibers for wearable electronics. Advanced Functional Materials. 2015;**25**(21):3114-3121

[113] Chen S et al. Conductive, tough, hydrophilic poly (vinyl alcohol)/graphene hybrid fibers for wearable supercapacitors. Journal of Power Sources. 2016;**319**:271-280

[114] Seyedin S et al. Knitted strain sensor textiles of highly conductive all-polymeric fibers. ACS Applied Materials & Interfaces. 2015;**7**(38):21150-21158

[115] Neves AI et al. Towards conductive textiles: Coating polymeric fibres with graphene. Scientific Reports. 2017;**7**(1):4250

[116] Rayhan MG et al. Conductive textiles for signal sensing and technical applications. Signals. 2023;**4**:1-39. DOI: 10.3390/signals4010001

[117] Lee S et al. Stretchable electronics: Ag nanowire reinforced highly stretchable conductive fibers for wearable electronics (adv. Funct. Mater. 21/2015). Advanced Functional Materials. 2015;**25**(21):3105-3105

[118] Suat C, Nilsen SE. The electrical and physical effects of yarns containing metal

wire on knitted fabric. Annals of the University of Oradea. Fascicle of Textiles, Leatherwork. 2019;**20**(1):2457-4880

[119] Hossain MA, Rahman M. A review of nano particle usage on textile material against ultra violet radiation. Journal of Textile Science and Technology. 2015;**1**(03):93

[120] El-Naggar ME, Shaarawy S, Hebeish A. Bactericidal finishing of loomstate, scoured and bleached cotton fibres via sustainable in-situ synthesis of silver nanoparticles. International Journal of Biological Macromolecules. 2018;**106**:1192-1202

[121] Hassabo AG et al. Development of multifunctional modified cotton fabric with tri-component nanoparticles of silver, copper and zinc oxide. Carbohydrate Polymers. 2019;**210**:144-156

[122] Fangueiro R et al. Wicking behavior and drying capability of functional knitted fabrics. Textile Research Journal. 2010;**80**(15):1522-1530

[123] Parada M. Imaging and Modeling of Wicking in Textiles. Zurich, Switzerland: ETH Zurich; 2018

[124] Kim H-S, Michielsen S, DenHartog E. Wicking in textiles at rates comparable to human sweating. Colloids and Surfaces A: Physicochemical and Engineering Aspects. 2021;**622**:126726

[125] Yetisen AK et al. Nanotechnology in textiles. ACS Nano. 2016;**10**(3):3042-3068

[126] Liu Y, Xin JH, Choi C-H. Cotton fabrics with single-faced superhydrophobicity. Langmuir. 2012;**28**(50):17426-17434

[127] Babar AA et al. Tailoring differential moisture transfer performance of nonwoven/polyacrylonitrile-SiO2 nanofiber composite membranes. Advanced Materials Interfaces. 2017;**4**(15):1700062

[128] Zaman M et al. Hydrophilic modification of polyester fabric by applying nanocrystalline cellulose containing surface finish. Carbohydrate Polymers. 2013;**91**(2):560-567

[129] Gocek İ, Duru SC. Investigating the effects of wicking and antibacterial finishing treatments on some comfort characteristics of Meryl skinlife for seamless activewear/sportswear. Journal of Engineered Fibers and Fabrics. 2019;**14**:1558925019852790

[130] Hezavehi E, Shahidi S, Zolgharnein P. Effect of dyeing on wrinkle properties of cotton cross-linked by butane tetracarboxylic acid (BTCA) in presence of titanium dioxide (TiO2) nanoparticles. Autex Research Journal. 2015;**15**(2):104-111

[131] Lam Y, Kan CW, Yuen C. Effect of concentration of titanium dioxide acting as catalyst or co-catalyst on the wrinkle-resistant finishing of cotton fabric. Fibers and Polymers. 2010;**11**:551-558

[132] Gong C et al. Thermosensitive polymeric hydrogels as drug delivery systems. Current Medicinal Chemistry. 2013;**20**(1):79-94

[133] Raza ZA, Anwar F. Fabrication of chitosan nanoparticles and multi-response optimization in their application on cotton fabric by using a Taguchi approach. Nano-Structures & Nano-Objects. 2017;**10**:80-90

Chapter 8

Thermodynamic and Kinetic Stability of Cosmetic Nanoemulsions

Omolade Ajayi

Abstract

In designing cosmetics and personal care products, the evolution of nanoemulsions has catalyzed groundbreaking advancements in enhancing the efficiency and delivery of actives aimed at rejuvenating and fortifying skin health. The ability of nanoemulsions to encapsulate both hydrophilic and hydrophobic actives while navigating biological barriers ensures their precise and targeted release, enhancing therapeutic efficacy while mitigating adverse effects. This chapter aims to elucidate the pivotal role of structural stability on the integrity, shelf-life, and functionality of these formulations, thereby driving innovation and progress in the formulation and design of highly effective cosmetics. By exploring the fundamental principles governing stability within these formulations, this chapter seeks to elucidate the critical factors shaping their design and longevity, ultimately paving the way for innovative advancements in the field.

Keywords: nanoemulsions, thermodynamic stability, kinetic stability, cosmetics, pharmaceuticals, skin care, formulation stability

1. Introduction

Emulsions are commonly used in cosmetics in the form of lotions and creams, and are in fact, the most popular form of cosmetic products [1]. They are composed of two (or more) immiscible liquid phases existing in a colloidal system, where one phase (the internal/dispersed phase) is uniformly suspended as fine droplets within the matrix of the other (the external/continuous phase). Most emulsions are bi-phased and contain one oil phase and one water phase. Depending on the nature of the dispersed and continuous phase, bi-phase emulsions are usually classified as either water-in-oil (W/O) or oil-in-water (O/W). In these systems, the "oil phase" refers to a non-polar substance or a mixture of hydrophobic materials such as plant oils, waxes, silicones, fatty acids, fatty alcohols and so on, that are essentially immiscible with water, while the "water phase" refers to water as a sole ingredient or an aqueous solution of polar molecules that are readily soluble in water. With recent advancements in emulsion technology, multiple emulsions can also be formulated in which the droplet of the dispersed phase is an emulsion itself and contains smaller dispersed droplets of another liquid phase [2]. Typically, these are referred to as double emulsions and could be W/O/W, O/W/O, W/O/O, O/W/W, O/O/W, or W/W/O (where W is water and O is oil) [3]. Emulsions can also be classified based on the droplet size of the dispersed phase. Under these classifications, the dispersed phase of nanoemulsions have extremely fine droplets of radius ranging from 20 to 200 nm and a narrow droplet size distribution [1, 4–6]. Under this classification, other types of emulsions known as microemulsions and macroemulsions exist and these have droplets sizes larger than that of nanoemulsions. This chapter would be focusing on nanoemulsions as they have been pivotal to advancement in the design and formulation of cosmetic products with enhanced efficacy.

As contemporary living standards progress and an aging population becomes more prevalent, a growing number of individuals are developing a keen interest in the subjects of aging, and skin care products with anti-aging actives have become even more appealing [7–11]. For example, due to their specific amino acid sequences, peptides exhibit the potential to stimulate collagen production and cellular regeneration, thereby offering promising anti-aging and skin-repair properties. Nanotechnology has been shown to be an effective strategy for improving the bioavailability and biostability of peptides [12].

Another class of actives derived from vitamin A, known as retinoids (such as retinol, retinyl acetate, retinyl propionate, retinyl palmitate and tretinoin), are particularly effective at slowing the skin aging process. However, these actives have been demonstrated to induce irritations, burning sensations, and even skin diseases such as dermatitis [13]. Additionally, they are known to be chemically unstable, being easily oxidized, and isomerized by heat, oxygen, and light due to their reactive conjugated double bonds [14, 15]. Nanoemulsions have been shown to improve the stability and reduce the irritations caused by retinoids [16–18]. Yang et al. developed nanoemulsions loaded with retinyl propionate and these were shown to exhibit highly enhanced transdermal delivery of the active compared to conventional emulsions. The nanoemulsions had droplet size <100 nm and were highly stable with a high retention rate exceeding 80% after storage at elevated temperature (50°C) for 30 days [18].

Asides anti-aging benefits, skin care products are also essential to supporting one of the skins most important function, which is to serve as a robust barrier against trans epidermal water loss (TEWL) [19]. The effectiveness of this permeability barrier is associated with the function of the stratum corneum (SC) lipid lamellae, which forms the intercellular layer of the skin with a distinct and diverse composition compared to the lipids found in biological membranes, 50% of which is a hydrophobic substance known as ceramides [20–24]. Depletion in the skin's natural ceramide content, corresponds with observable cutaneous signs such as dryness, diminished elasticity, and increased roughness and even skin diseases such as atopic dermatitis and psoriasis [25, 26]. In recent years, synthetic ceramides are increasingly being used in cosmetic formulations to support the skin barrier, maintain skin hydration, and protect the skin against potential damage [25, 27–29]. However, due to their high molecular weight and extreme hydrophobicity, traversing the layers of the skin is a challenging task for these ingredients [2, 30]. Tashiro et al. [31] showed that nano focus technology could be pivotal to the formulation of ceramide based cosmetic products. They developed a human-type nano-ceramide dispersion with droplet size of about 20 nm for the purpose of improving ceramide permeation and it performed more than 9 times better than conventional formulations. Guzmán et al. [32] also prepared stable nanoemulsions (droplet size of about 200 nm) for the enhanced penetration of a highly hydrophobic ceramide-like molecule ((9Z)-N-(1,3-dihydroxyoctadecan-2-yl)octadec-9-enamide). The emulsions showed good long-term stability for several weeks. Their findings indicated that nanoemulsions can be used as a highly promising tool for improving the bioavailability of hydrophobic molecules and consequently the efficacy of cosmetic formulations in which they are encapsulated.

Another application of cosmetic nanoemulsions is in the topical administration of epidermal growth factors which are cosmetic actives know to promote cellular growth, proliferation, and differentiation beneficial for accelerated wound healing and tissue regeneration. Due to their restricted percutaneous absorption, inconsistent efficacy results have been obtained from the use of growth factors in cosmetic products. To address this, Sang et al. [33] developed a low-molecular-weight protamine conjugated nanoemulsion-dispersed hydrogel, loaded with three growth factors. The hydrogels showed significantly higher skin permeation than native formulations.

Research indicates that fundamentally, the effective transdermal delivery and stability of these actives are two major factors that affect their application in cosmetic formulation design.

2. Emulsion stability

Stability is the ability of a system to resist physical or chemical changes and remain intact under varying conditions [34]. In chemistry, there are two main types of stability: thermodynamic stability and kinetic stability. Thermodynamic stability refers to the tendency of a system to reach and maintain a state of minimum energy at equilibrium. In thermodynamics, a stable system is one that has achieved the lowest possible free energy and a reaction that produces a thermodynamically stable product ensures that the overall free energy of the system is decreased, and the system tends to remain in this state over time. In addition, the second law of thermodynamics for irreversible phenomena states that "In an irreversible or spontaneous change from one equilibrium state to another (for example, the equalization of temperature of two bodies A & B, when brought in contact), the entropy always increases" [35]. This means that a reaction that produces a stable system ensures that the system reaches the highest possible entropy under given conditions. Combining both conditions, a thermodynamically stable system must be in a favorable state in terms of free energy and entropy and its potential of undergoing spontaneous changes are minimized.

Kinetic stability on the other hand, refers to the resistance of a system to undergoing transformation or reaction over time [36] because of the presence of a barrier to a specific reaction rather than the thermodynamic favorability of the final state. Systems with high kinetic stability possess a substantial high kinetic barrier, hindering the occurrence of spontaneous reactions (or destabilization) even in cases where a destabilizing reaction is thermodynamically favorable (meaning the system is not thermodynamically resistant to this destabilizing reaction under the conditions of such reaction) [37]. Hence a system can exhibit various stability combinations: it may be both thermodynamically and kinetically stable, it can be thermodynamically unstable yet kinetically stable, it can be thermodynamically stable but kinetically unstable, or it can be both thermodynamically and kinetically unstable. Since emulsions are comprised of two phases that are inherently immiscible, combining these two phases into a stable mixture requires work to ensure either thermodynamic stability or kinetic stability or both.

The stability of nanoemulsions directly influence their functionality in cosmetic and pharmaceutical applications. For instance, in cosmetics, stable nanoemulsions allow for controlled release of active compounds like peptides and epidermal growth factors, ensuring optimal penetration into the skin for desired effects. Achieving and maintaining stability is fundamental to ensuring the efficacy of active compounds throughout the shelf life and lifespan of these innovative formulations.

2.1 Thermodynamic stability of nanoemulsions

Consider **Figure 1** where the two phases (oil and water) of an emulsion come in contact.

The two phases are immiscible because the cohesive forces between the molecules of either phase is greater than the adhesive forces between the two phases. Each phase has uniform thermodynamic properties (both intensive and extensive) that are different from the other phase. So, when both phases are brought together, there is repulsion between the molecules of both phases, and this creates tension at the contact point of both liquids. To maintain thermodynamic stability, both liquids will tend of maintain a small contact area as much as possible because the tension created is associated with a concomitant increase in the free energy of the system [38]. This contact area is also known as an interface and Gibbs likened this interface to a dividing line [39]. The interfacial tension at this dividing line can be derived mathematically from the second law of thermodynamics. Given the Gibbs free energy as:

$$G = U + pV - TS \tag{1}$$

Where U is the internal energy (J), p is pressure (pa), V is volume (m^3), T is absolute temperature (K), and S is the entropy (J/K).

Taking the oil phase as the dispersed phase, the emulsion formed is an o/w emulsion. With emulsification, the system changes from two separated phases to one emulsified bulk with the droplets of the oil dispersed within the matrix of the water (**Figure 2**) and multiple interfaces are formed around the oil droplets.

As more droplets are formed, the total interfacial surface area is increased, and the free energy of the system also increases. The change in Gibbs free energy as the system changes from two separate phases to emulsified at constant temperature is given as:

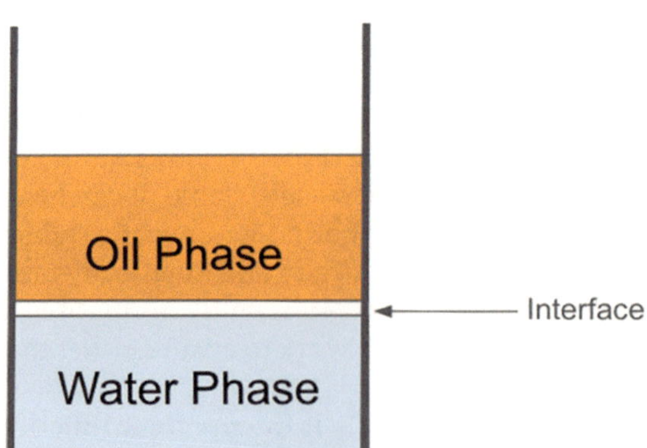

Figure 1.
Schematic diagram showing the interface formed at the contact of two immiscible liquids (oil and water).

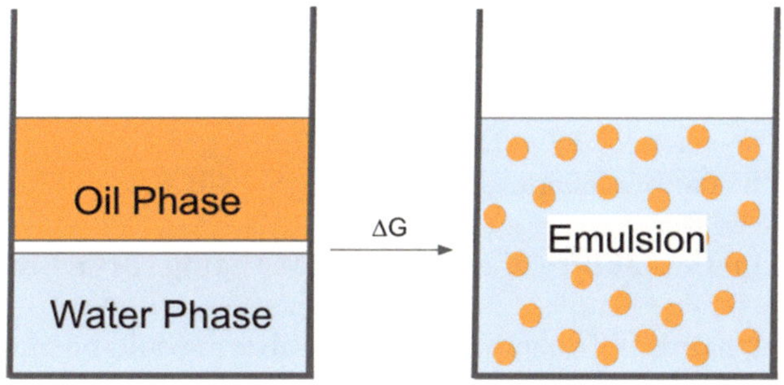

Figure 2.
Schematic diagram showing an emulsion formed from the oil phase droplets uniformly dispersed within the water phase.

$$\Delta G = \Delta U + p\Delta V + V\Delta p - T\Delta S \quad (2)$$

The entropy (ΔS) is a measure of the extent of disorder in a system which is a measure of the extent of reduction in droplet size of the dispersed phase (or an increase in the number of droplets). At constant pressure and composition, the change in Gibbs free energy of the system is given as:

$$\Delta G = (\Delta U + p\Delta V) - T\Delta S = \Delta H - T\Delta S \quad (3)$$

Where ΔH is the change in enthalpy of the system. If the volume of the system remains constant as the emulsion is formed (which is expected), the enthalpy of the system is equal to the change in internal energy of the system which is equal to the work (ΔW) done on the system to form the emulsion.

$$\Delta G_{form} = \Delta W - T\Delta S \quad (4)$$

This work can be expressed as a function of the interfacial tension (γ) and the change in total surface area (ΔA) of the interface as the emulsion is formed

$$\Delta W = \gamma \Delta A \quad (5)$$

Therefore,

$$\Delta G_{form} = \gamma \Delta A - T\Delta S \quad (6)$$

As the droplet size of the dispersed phase decreases, the total interfacial surface area increases, and the energy input increases ($\gamma\Delta A$) . In the same way, the configuration entropyterm (T ΔS) increases because the dispersion of the oil droplets is accompanied by an increase in disorderliness within the system. Hence, an increase in entropy (favorable in terms of thermodynamic stability). However, with increasing entropy, the configuration entropy term in Eq. (6) becomes increasingly negative (hence thermodynamically favorable). The total change in Gibbs free energy of formation (ΔG_{form}) is hence higher than the free energy of the separated phases and the formation of the nanoemulsion is thermodynamically unfavorable [5].

2.2 Controlling kinetic stability of nanoemulsions using surfactants

Being thermodynamically unstable, a nanoemulsion would breakdown over time. As shown in **Figure 3**, the rate of breakdown is dependent on the height of the activation energy between the nanoemulsion and the separated phase.

Since the interfacial free energy change with reduction in droplet radius dominates in nanoemulsion formation, stability of the emulsion can be achieved with reduction in interfacial tension. This can be achieved with the addition of one or more surface active agent (surfactants) which help to reduce the interfacial tension and form an energy barrier between the phases [40, 41]. Ideally, with the use of surfactants, an ultra-low interfacial tension, typically below 10^{-3} mN/m can be obtained [42]. In addition to lowering interfacial tension, surfactants also contribute to kinetic stabilization by creating a complex electric film on the surfaces of dispersed phase droplets [1]. This energy barrier combined with the very small droplet size of nanoemulsions make them very kinetically stable compared to other types of emulsions [43]. This makes them very resistant to destabilization processes such as coalescence and flocculation, but they are susceptible to Ostwald ripening over time [4, 44–48]. By carefully selecting the particle size distribution, oil type, and emulsifier type, nanoemulsions

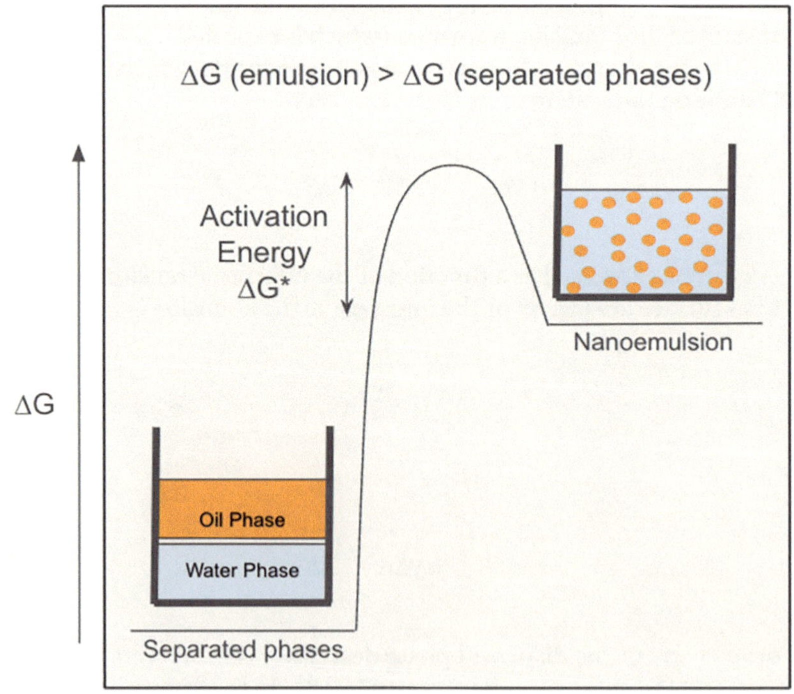

Figure 3.
Schematic representation of the free energy of formation of a nanoemulsion.

with extended kinetic stabilities lasting months or even years can be formulated. Additionally, employing an oil phase with minimal water solubility aids in preventing droplet growth through Ostwald ripening [34, 49–51].

Surfactants are a large class of chemicals but those used specially for emulsification processes are known as emulsifiers and the type of emulsifier chosen for a specific nanoemulsion system depends on how lipophilic the oil phase of the emulsion is. This is typically expressed in terms of its hydrophile-lipophile balance (HLB) number; a number determined by calculating the percentages of molecular weights for the lipophilic and hydrophilic portions of the surfactant molecule. The concept was originally developed by Griffin in 1949 but numerous empirical formulas have been developed to calculate the HLB of various surfactants based on the molecular structure of different surfactants [40, 52–55]. Analytical experimental methods such as gas chromatography and NMR spectroscopy have also been developed to determine the HLB of surfactants [53, 56–60]. To obtain a stable emulsion, the HLB of the emulsifier must be approximately equal to the HLB of the oil phase and the mechanism for stabilizing the nanoemulsion relies on the chemical structure of the surfactant used.

In the Retinyl propionate loaded nanoemulsions developed by Yang et al. [18], polysorbate-20 was chosen as the main emulsifier and was used at 5%. Positively charged behentrimonium chloride was used at 1% as a co-surfactant to form an electric layer around the droplets and varying concentrations of hydrophobically modified Inulin was used to increase the viscosity of the water phase ensuring that the oil phase droplets were densely packed. The superior stabilization property exhibited by this nanoemulsion system was attributed to the densely-packed, sterically-hindered and highly-charged interfacial layers achieved using multiple emulsifiers at the oil-water interface [18].

The long-term stability of nano-emulsion, governed majorly by kinetics, make them spectacularly appealing for cosmetic application and even though they are inherently thermodynamically unstable, they are sometimes referred to as 'Approaching Thermodynamic Stability' [48]. Also, even though the droplet size of nanoemulsions contributes to its desirable functionality, the stability of nanoemulsions is not exactly a function of how small the droplet size of the dispersed is. Bernardi et al. [6] conducted thermal stress tests on five O/W nanoemulsions with droplet sizes ranging from 45 nm to 303 nm and observed that the formulation with the lowest droplet size (45 nm ± 12) was the only one that showed signs of instability. The formulation was composed of 10% rice bran oil, 20% surfactant blend (Sorbitan Oleate/PEG-30 castor oil) and 70% water. Hence, irrespective of the amount of surfactant used, the formulation method chosen, or the droplet size achieved, it is important to conduct extensive long term stability studies on nanoemulsions to establish the product's effectiveness, shelf life and structural integrity over time.

3. Methods of formulating nanoemulsions

In addition to using surfactants to ensure kinetic stability of nanoemulsions, formulation methods to ensure the droplet of the dispersed phase are small enough and uniform within the matrix of the continuous phase is very crucial to maintaining long-term stability of nanoemulsions. Due to their thermodynamic instability, nanoemulsions cannot be formed spontaneously [48]. External energy input is required to transform the two separate phases into a colloidal dispersion. At a minimum, energy input surpassing the positive free energy related to increasing the contact area between the oil and water phases ($\Delta G = \gamma \Delta A$) must be applied. Nanoemulsion fabrication methods can be broadly classified as either high-energy or low-energy approaches, depending on the underlying physicochemical mechanism of droplet disruption.

3.1 High energy methods

3.1.1 Ultra-sonification

Ultrasonication, a prominent high-energy method in nanoemulsion formulation, harnesses the power of high-frequency sound waves to create fine and uniform droplets. During ultrasonication, the energy generated by ultrasonic waves induces cavitation, creating microbubbles in the liquid [61]. The subsequent collapse of these bubbles generates intense shear forces, breaking down larger droplets into smaller nanoscale droplets. This technique is particularly effective for both oil-in-water and water-in-oil nanoemulsions as it proves to be particularly valuable for formulating drug delivery systems, improving solubility, and enhancing the therapeutic efficacy of poorly water-soluble compounds [62]. The key advantage of ultrasonication lies in its ability to produce nanoemulsions with a narrow droplet size distribution, enhancing stability and bioavailability. It allows for precise control over particle size, and the intensity and duration of ultrasonication can be adjusted to achieve the desired characteristics for a specific application [63].

3.1.2 High pressure homogenization

High-pressure homogenization is a robust method extensively employed in the formulation of nanoemulsions, offering precise control over droplet size and distribution [64]. This technique involves subjecting the emulsion mixture to elevated pressures (as high as 350 MPa), typically achieved by forcing it through a narrow valve or aperture [65]. The intense shearing forces generated during this process result in the reduction of larger droplets into smaller, nanoscale droplets. The process begins with the preparation of an emulsion containing oil, water, and surfactants. As the emulsion passes through the homogenization apparatus, the narrow valve or aperture generates turbulent forces, breaking down larger droplets into smaller ones [65]. The pressure applied ensures uniformity in droplet size, contributing to the stability and quality of the resulting nanoemulsion. Despite its efficacy, it's crucial to consider the potential impact of high-pressure homogenization on the stability of sensitive compounds. The process may generate heat as the formulation is exposed to high pressures, and careful temperature control and pressure adjustment would be required to preserve the integrity of bioactive ingredients. In summary, high-pressure homogenization stands as a versatile and efficient method for nanoemulsion formulation, providing a controlled environment to create stable and finely dispersed droplets.

3.1.3 Microfluidization

Microfluidization is a sophisticated patented technology employed in the formulation of nanoemulsions [66]. It uses the same principles of high-pressure homogenization but it can be used to prepare nanoemulsions with high oil content (up to 50% oil) [67–69]. It involves forcing the emulsion through small channels or chambers at high velocities, resulting in intense shearing forces and droplet size reduction. Microfluidization relies on the generation of high shear forces within the microchannels or chambers. As the emulsion passes through these confined spaces at elevated pressures, intense shearing occurs, breaking down larger droplets into smaller nanoscale droplets. The pressure applied during microfluidization is a critical parameter. It can be precisely controlled to achieve the desired droplet size and distribution, making this method suitable for producing nanoemulsions with specific characteristics. The initial step in this process involves preparing an emulsion with oil, water, and suitable surfactants or stabilizers. The emulsion is then forced through the microfluidization equipment, typically composed of chambers with interaction zones where high-pressure fluid dynamics occur. The intense shear forces and microscale turbulence lead to the breakup of larger droplets, resulting in the formation of a nanoemulsion with a narrow droplet size distribution [67–69].

While both microfluidization and high-pressure homogenization are similar methods employed for particle size reduction and emulsification, they differ in their underlying principles, equipment, and specific applications. Microfluidization relies on intense shearing forces generated by the microscale turbulence within small channels or chambers while high-pressure homogenization involves forcing the emulsion through a narrow valve or aperture at high pressures, creating shearing forces that break down larger droplets into smaller ones. While microfluidization and high-pressure homogenization share similarities, the specific design of the equipment, the mechanisms of shearing, and the level of control over droplet size can vary between the two methods. Also, microfluidization is considered energy-efficient compared to other high-energy methods. The process generates less heat, which can be crucial for preserving the stability of sensitive compounds. However, the need for a coarse emulsion limits the scalability of microfluidization [70].

The choice between microfluidization and high-pressure homogenization often depends on the specific requirements of the nanoemulsion formulation, and the characteristics desired for the final product.

3.2 Low energy methods

3.2.1 High speed homogenization

High-speed homogenization is a very simple and widely employed method for the fabrication of nanoemulsions, offering a robust and efficient approach to produce finely dispersed and stable colloidal systems. In this method, a high shear force, usually around 6000–24,000 rpm [71], is applied to the emulsion system by a high-speed rotor stator [71–73]. During the process, the intense mechanical forces generated by the rotor cause disruption and breakup of the larger droplets present in the emulsion. With elevated homogenization speeds, the solution experiences a greater energy density, leading to a direct reduction in emulsion droplet size. Shear stress exhibits an inverse relationship with emulsion droplet size, meaning that an increase in shear stress results in a reduction in droplet size, ultimately producing the nanoemulsion [72]. The rapid and repeated shearing action helps to achieve a more uniform distribution of the components, leading to the creation of a stable nanoemulsion.

Parameters such as homogenization speed, duration, and formulation composition can be optimized to tailor the properties of the resulting nanoemulsion, making this method versatile enough for both laboratory and industrial scale. The efficiency and scalability of high-speed homogenization contributes to its widespread adoption as a reliable technique for nanoemulsion fabrication.

3.2.2 Phase inversion methods

Phase inversion methods represent a versatile approach in the creation of nanoemulsions, involving the manipulation of temperature and composition to induce transitions between oil-in-water (O/W) and water-in-oil (W/O) structures due to a spontaneous curvature of the surfactant [73, 74]. The two primary phase inversion methods are the Phase Inversion Temperature (PIT) method and the Phase Inversion Composition (PIC) method [74]. PIT relies on altering the temperature to induce phase transitions within the emulsion. By carefully adjusting the temperature, the system can switch between O/W and W/O structures, leading to the formation of nanoemulsions [59, 74]. PIC involves altering the composition of the dispersed phase to induce phase inversion and nanoemulsion formation [74, 75].

It's important to note that successful implementation of phase inversion methods requires a deep understanding of the system's components, their interactions, and the specific conditions needed to induce phase transitions. Additionally, considerations for the stability and shelf-life of the resulting nanoemulsions are crucial in practical applications. Overall, phase inversion methods contribute significantly to the versatility and adaptability of nanoemulsion technology in cosmetic formulation design.

Modarres-Gheisari et al. [70] did a comprehensive comparison of the different nanoemulsion fabrication methods and found that high speed homogenization had the highest usage percentage due to its relative simplicity.

4. Conclusion

Due to their extremely fine droplets and the presence of surfactants, nanoemulsions have relatively large interfacial surface area but low interfacial surface tension which has been shown to enhance the adsorption and controlled release of a vast range of ingredients in cosmetic products such as potent skin-protecting actives, antioxidants, rejuvenating plant extracts, UV filters as well as encapsulating volatile essential oils and aromatic compounds for alcohol free perfume formulations. Inherently, nanoemulsion formation is thermodynamically unfavorable but formulating stable nanoemulsions can be achieved with the use of one of more surfactants which act as a barrier against instability. This energy barrier imparts kinetic stability and is determined by factors like a proper balance of oil, water and surfactant content in the system, electrostatic repulsion between the oil droplets and steric hindrance provided by the surfactants.

Successful nanoemulsion formulation involves a balance between the choice of surfactants and co-surfactants, and processing method. Moreover, the intended application, whether in drug delivery through topical pharmaceutical products or in cosmetics influences the selection of the most appropriate formulation method. Continuous research and innovation in nanoemulsion technology continue to refine these methods, expanding their applications and enhancing the efficiency of nanoemulsion-based products for cosmetic applications.

Author details

Omolade Ajayi
Department of Chemical Engineering, Manhattan College, New York, USA

*Address all correspondence to: oajayi01@manhattan.edu

References

[1] Sikora E. Cosmetic emulsions. In: Monograph. Kraków: Cracow University of Technology Press; 2019. 118 p. ISBN/ISSN: 978-83-65991-50-8

[2] Nafisi. S, Maibach S. Chapter 22 - nanotechnology in cosmetics. In: Kazutami. S, Robert Lochhead R, Howard M, Yuji Y, editors. Cosmetic Science and Technology: Theoretical Principles and Applications. 1st ed. Amsterdam: Elsevier; 2017. p. 337-369. DOI: 10.1016/B978-0-12-802005-0.00022-7

[3] Hema S, Apaeajita K, Raunak-Kumar D, Priyanka S. Simple formulation and characterization of double emulsions variant designed to cary three bioactive agents. Heliyon: A Cell Press Journal. 2022;**8**(9):e10397. DOI: 10.1016/j.heliyon.2022.e10397

[4] Solans C, Izquierdo P, Nolla J, Azemar N, Garcia-Celma M. Nano-emulsions. Current Opinion in Colloid & Interface Science. 2005;**10**(3-4):102-110. ISSN 1359-0294. DOI: 10.1016/j.cocis.2005.06.004

[5] McClements DJ. Nanoemulsions versus microemulsions: Terminology, differences, and similarities. Soft Matter. 2013;**8**:1719-1729. DOI: 10.1039/C2SM06903B

[6] Bernardi D, Pereira T, Maciel N, Josiane B, Giselv V, Gustavo O, et al. Formation and stability of oil-in-water nanoemulsions containing rice bran oil: In vitro and in vivo assessments. Journal of Nanobiotechnology. 2011;**9**:44. DOI: 10.1186/1477-3155-9-44

[7] He X, Wan F, Su W, Xie W. Research Progress on skin aging and active ingredients. Molecules [Internet]. 2023;**28**(14):5556. Available from: https://www.mdpi.com/1420-3049/28/14/5556

[8] Bouchemal K, Briançon S, Perrier E, Fessi H. Nano-emulsion formulation using spontaneous emulsification: Solvent, oil and surfactant optimisation. International Journal of Pharmaceutics. 2004;**280**(1-2):241-251. DOI: 10.1016/j.ijpharm.2004.05.016

[9] Romes NB, Abdul Wahab R, Abdul HM. The role of bioactive phytoconstituents-loaded nanoemulsions for skin improvement: A review. Biotechnology & Biotechnological Equipment. 2021;**35**(1):711-729

[10] Sun R, Xia N, Xia Q. Non-aqueous nanoemulsions as a new strategy for topical application of astaxanthin. Journal of Dispersion Science and Technology. 2019;**41**(12):1777-1788

[11] Ashaolu TJ. Nanoemulsions for health, food, and cosmetics: A review. Environmental Chemistry Letters. 2021;**19**(4):3381-3395. DOI: 10.1007/s10311-021-01216-9. Epub 2021 Mar 15. PMID: 33746662; PMCID: PMC7956871

[12] Zhang X, Li X, Zhao Y, Zheng Q, Wu Q, Yu Y. Nanocarrier system: An emerging strategy for bioactive peptide delivery. Frontiers in Nutrition. 2022;**9**. DOI: 10.3389/fnut.2022.1050647

[13] Mukherjee S, Date A, Patravale V, Korting HC, Roeder A, Weindl G. Retinoids in the treatment of skin aging: An overview of clinical efficacy and safety. Clinical Interventions in Aging. 2006;**1**(4):327-348

[14] Bempong DK, Honigberg IL, Meltzer NM. Normal phase LC-MS determination of retinoic acid

degradation products. Journal of Pharmaceutical and Biomedical Analysis. 1995;**13**(3):285-291

[15] McCaffery P, Evans J, Koul O, Volpert A, Reid K, Ullman MD. Retinoid quantification by HPLC/MSn. Journal of Lipid Research. 2002;**43**(7):1143-1149

[16] Oh YK, Kim MY, Shin JY, Kim TW, Yun MO, Yang SJ, et al. Skin permeation of retinol in Tween 20-based deformable liposomes: In-vitro evaluation in human skin and keratinocyte models. Journal of Pharmacy and Pharmacology. 2006;**58**(2):161-166

[17] Park H, Mun S, Kim YR. UV and storage stability of retinol contained in oil-in-water nanoemulsions. Food Chemistry. 2019;**272**:404-410. DOI: 10.1016/j.foodchem.2018.08.076

[18] Yang Y, Yan S, Yu B, Gao C, Wang J. Hydrophobically modified inulin based nanoemulsions for enhanced stability and transdermal delivery of retinyl propionate. Colloids and Surfaces A: Physicochemical and Engineering Aspects. 2022;**653**:129883-129883

[19] Harding CR. The stratum corneum: Structure and function in health and disease. Dermatologic Therapy. 2004;**17**(s1):6-15

[20] Sajić D, Asiniwasis R, Skotnicki-Grant S. A look at epidermal barrier function in atopic dermatitis: Physiologic lipid replacement and the role of ceramides. PubMed. 2012;**17**(7):6-9

[21] The important role of stratum corneum lipids for the cutaneous barrier function. Biochimica et Biophysica Acta (BBA) - Molecular and Cell Biology of Lipids. 2014;**1841**(3):295-313. DOI: 10.1016/j.bbalip.2013.11.006

[22] Moore DJ, Rawlings AV. The chemistry, function and (patho) physiology of stratum corneum barrier ceramides. International Journal of Cosmetic Science. 2017;**39**(4):366-372. DOI: 10.1111/ics.12399

[23] Feingold KR. Thematic review series: Skin lipids. The role of epidermal lipids in cutaneous permeability barrier homeostasis: Fig. 1. Journal of Lipid Research. 2007;**48**(12):2531-2546

[24] Elias PM. Epidermal barrier function: Intercellular lamellar lipid structures, origin, composition and metabolism. Journal of Controlled Release. 1991;**15**(3):199-208. DOI: 10.1016/0168-3659(91)90111-P

[25] Tessema EN, Gebre-Mariam T, Neubert RHH, Wohlrab J. Potential applications of phyto-derived ceramides in improving epidermal barrier function. Skin Pharmacology and Physiology. 2017;**30**(3):115-138

[26] Imokawa G, Abe A, Jin K, Higaki Y, Kawashima M, Hidano A. Decreased level of ceramides in stratum corneum of atopic dermatitis: An etiologic factor in atopic dry skin? Journal of Investigative Dermatology. 1991;**96**(4):523-526

[27] Pons-Guiraud A. Dry skin in dermatology: A complex physiopathology. Journal of the European Academy of Dermatology and Venereology. 2007;**21**(s2):1-4. DOI: 10.1111/j.1468-3083.2007.02379.x

[28] Rogers J, Harding C, Mayo A, Banks J, Rawlings A. Stratum corneum lipids: The effect of ageing and the seasons. Archives of Dermatological Research. 1996;**288**(12):765-770. DOI: 10.1007/BF02505294

[29] Spada F, Barnes TM, Greive KA. Skin hydration is significantly increased by a cream formulated to mimic the skin's own natural moisturizing systems. Clinical, Cosmetic and Investigational

Dermatology. 2018;**11**:491-497. DOI: 10.2147/CCID.S177697

[30] Alonso A, Goñi FM. The physical properties of ceramides in membranes. Annual Review of Biophysics. 2018;**47**:633-654. DOI: 10.1146/annurev-biophys-070317-033309

[31] Tashiro T, Nakaune A, Kosugi T, Arakawa J, Mori H, Serizawa S, et al. Development of functional cosmetics "ASTALIFT JELLY AQUARYSTA". Fujifilm Research & Development. 2011;**56**:1-4

[32] Guzmán E, Fernández-Peña L, Rossi L, Bouvier M, Ortega F, Rubio RG. Nanoemulsions for the encapsulation of hydrophobic actives. Cosmetics. 2021;**8**(2):45. DOI: 10.3390/cosmetics8020045

[33] Choi SW, Pangeni R, Park JW. Nanoemulsion-based hydrogel for topical delivery of highly skin-permeable growth factor combinations: Preparation and in vitro evaluation. Journal of Nanoscience and Nanotechnology. 2017;**17**(4):2363-2369. DOI: 10.1166/jnn.2017.13318

[34] McClements DJ. Food Emulsions: Principles, Practices, and Techniques. 2nd ed. Boca Raton: CRC Press; 2004. 632 p. DOI: 10.1201/9781420039436

[35] Uhlenbeck G, Ford G. Lectures in Statistical Mechanics. Journal of Applied Mathematics and Mechanics. 1963;**43**(10):520. DOI: 10.1002/zamm.19630431044

[36] Aihara J-i. Reduced HOMO-LUMO gap as an index of kinetic stability for polycyclic aromatic hydrocarbons. Journal of Physical Chemistry. 1999;**103**:7487-7495

[37] Pross A. Stability in chemistry and biology: Life as a kinetic state of matter. Pure and Applied Chemistry Journal. 2005;**77**(11):1905-1921. DOI: 10.1351/pac200577111905

[38] Lautrup B. Physics of Continuous Matter: Exotic and Everyday Phenomena in the Macroscopic World. 2nd ed. Boca Raton, FL: A Taylor & Francis Book (CRC Press); 2019

[39] Gibbs JW. Collected Papers of J Willard Gibbs: Thermodynamics. New York, USA: Dover Publications; 1961

[40] Becher P. Encyclopedia of Emulsion Technology. New York: Marcel Dekker; 1996

[41] Adams F, Binks BP, Walstra P, Brooks BW, Richmond HN, Zerfa M, et al. Modern Aspects of Emulsion Science. Cambridge: Royal Society of Chemistry; 2007

[42] Devarajan V, Ravichandran V. Nanoemulsions: As modified drug delivery tool. International Journal of Comprehensive Pharmacy. 2011;**2**

[43] Anton N, Vandamme TF. Nano-emulsions and micro-emulsions: Clarifications of the critical differences. Pharmaceutical Research. 2011;**28**(5):978-985

[44] Tadros TF. Nanodispersions. Berlin, Boston: Walter de Gruyter GmbH & Co KG; 2015

[45] Anton N, Vandamme TF. The universality of low-energy nano-emulsification. International Journal of Pharmaceutics. 2009;**377**(1-2):142-147

[46] Anton N, Benoit JP, Saulnier P. Design and production of nanoparticles formulated from nano-emulsion templates—A review. Journal of Controlled Release. 2008;**128**(3):185-199

[47] Izquierdo P, Esquena J, Tadros TF, Dederen C, Garcia MJ, Azemar N, et al. Formation and stability of nano-emulsions prepared using the phase

inversion temperature method. Langmuir. 2001;**18**(1):26-30

[48] Tadros T, Izquierdo P, Esquena J, Solans C. Formation and stability of nano-emulsions. Advances in Colloid and Interface Science. 2004;**108-109**:303-318. DOI: 10.1016/j.cis.2003.10.023

[49] McClements DJ, Rao J. Food-grade nanoemulsions: Formulation, fabrication, properties, performance, biological fate, and potential toxicity. Critical Reviews in Food Science and Nutrition. 2011;**51**(4):285-330. DOI: 10.1080/10408398.2011.559558

[50] McClements DJ. Edible nanoemulsions: Fabrication, properties, and functional performance. Soft Matter. 2011;**7**(6):2297-2316. DOI: 10.1039/C0SM00549E

[51] Kruglyakov PM. Hydrophile - Lipophile Balance of Surfactants and Solid Particles. Penza, Russia: Elsevier; 2000

[52] Griffin WC. Classification of surface-active agents by "HLB". Journal of Cosmetic Science. 1949;**1**:311-326

[53] Griffin WC. Calculation of HLB values of non-ionic surfactants. Journal of the Society of Cosmetic Chemists. 1954;**5**:249-256

[54] Davies JT. A quantitative kinetic theory of emulsion type. I physical chemistry of the emulsifying agent. In: Gas/Liquid and Liquid/Liquid Interfaces. Proceedings of the 2nd International Congress Surface Activity. London: Butterworths; 1957. pp. 426-438

[55] Vaughan CD, Rice DA. Predicting O/W emulsion stability by the "required HLB equation". Journal of Dispersion Science and Technology. 1990;**11**(1):83-91

[56] Rabaron A, Cavé G, Puisieux F, Seiller M. Physical methods for measurement of the HLB of ether and ester non-ionic surface-active agents: H-NMR and dielectric constant. International Journal of Pharmaceutics. 1993;**99**(1):29-36

[57] Badwan AA, Cham TM, James KC, Pugh WJ. A relationship between foaming and HLB value. Journal of Pharmacy and Pharmacology. Feb 1980;**2**(1):45-51. DOI: 10.1111/j.1467-2494.1980.tb00233.x. PMID: 19467079

[58] Marszall L. Cloud point and emulsion inversion point. Fette, Seifen, Anstrichmittel. 1977;**79**(1):41-44. DOI: 10.1002/lipi.19770790107

[59] Hayes ME, El-Emary M, Schechter RS, Wade WH. The relation between the EACNmin concept and surfactant HLB. Journal of Colloid and Interface Science. 1979;**68**(3):591-592. DOI: 10.1016/0021-9797(79)90314-x

[60] Gogate PR, Kabadi AM. A review of applications of cavitation in biochemical engineering/biotechnology. Biochemical Engineering Journal. 2009;**44**(1):60-72. DOI: 10.1016/j.bej.2008.10.006

[61] Abiraman T, Ramanathan E, Kavitha G, Rengasamy R, Balasubramanian S. Synthesis of chitosan capped copper oxide nanoleaves using high intensity (30 kHz) ultrasound sonication and their application in antifouling coatings. Ultrasonics Sonochemistry. 2017;**34**:781-791. DOI: 10.1016/j.ultsonch.2016.07.013

[62] Maali A, Mosavian MTH. Preparation and application of nanoemulsions in the last decade (2000-2010). Journal of Dispersion Science and Technology. 2013;**34**(1):92-105. DOI: 10.1080/01932691.2011.648498

[63] Yen CC, Chen YC, Wu MT, Wang CC, Wu YT. Nanoemulsion as a strategy for improving the oral bioavailability and anti-inflammatory activity of andrographolide. International Journal of Nanomedicine. 2018;**13**:669-680. DOI: 10.2147/IJN.S154824

[64] Floury J, Desrumaux A, Axelos MAV, Legrand J. Effect of high pressure homogenisation on methylcellulose as food emulsifier. Journal of Food Engineering. 2003;**58**(3):227-238

[65] Sutradhar KB, Amin ML. Nanoemulsions: Increasing possibilities in drug delivery. European Journal of Nanomedicine. 2013;**5**(2):97-110. DOI: 10.1515/ejnm-2013-0001

[66] Bai L, McClements DJ. Development of microfluidization methods for efficient production of concentrated nanoemulsions: Comparison of single- and dual-channel microfluidizers. Journal of Colloid and Interface Science. 2016;**466**:206-212. DOI: 10.1016/j.jcis.2015.12.039

[67] Ganesan P, Karthivashan G, Park SY, Kim J, Choi DK. Microfluidization trends in the development of nanodelivery systems and applications in chronic disease treatments. International Journal of Nanomedicine. 2018;**13**:6109-6121. DOI: 10.2147/IJN.S178077

[68] Luo X, Zhou Y, Bai L, Liu F, Deng Y, McClements DJ. Fabrication of β-carotene nanoemulsion-based delivery systems using dual-channel microfluidization: Physical and chemical stability. Journal of Colloid and Interface Science. 2017;**490**:328-335. DOI: 10.1016/j.jcis.2016.11.057

[69] Wilson RJ, Li Y, Yang G, Zhao CX. Nanoemulsions for drug delivery. Particuology [Internet]. 2021;**64**:85-97. Available from: https://www.sciencedirect.com/science/article/pii/S1674200121001176. DOI: 10.1016/j.partic.2021.05.009

[70] Modarres-Gheisari S, Gavagsaz-Ghoachani R, Malaki M, Safarpour P, Zandi M. Ultrasonic nano-emulsification – A review. Ultrasonics Sonochemistry. 2019;**52**:88-105. DOI: 10.1016/j.ultsonch.2018.11.005

[71] Mulia K, Safiera A, Pane IF, Krisanti EA. Effect of high speed homogenizer speed on particle size of polylactic acid. Journal of Physics: Conference Series. 2019;**1198**(6):062006. DOI: 10.1088/1742-6596/1198/6/062006

[72] Wang X, Jiang Y, Wang YW, Huang MT, Ho CT, Huang Q. Enhancing anti-inflammation activity of curcumin through O/W nanoemulsions. Food Chemistry. 2008;**108**(2):419-424. DOI: 10.1016/j.foodchem.2007.10.086

[73] Solé I, Pey CM, Maestro A, González C, Porras M, Solans C, et al. Nano-emulsions prepared by the phase inversion composition method: Preparation variables and scale up. Journal of Colloid and Interface Science. 2010;**344**(2):417-423

[74] Che Marzuki NH, Wahab RA, Abdul HM. An overview of nanoemulsion: Concepts of development and cosmeceutical applications. Biotechnology & Biotechnological Equipment. 2019;**33**(1):779-797. DOI: 10.1080/13102818.2019.1620124

[75] Shinoda K, Saito H. The effect of temperature on the phase equilibria and the types of dispersions of the ternary system composed of water, cyclohexane, and nonionic surfactant. Journal of Colloid and Interface Science. 1968;**26**(1):70-74

Chapter 9

The Nanoemulsion Technique, One of the Most Promising Strategies for Enhancing Drug Permeation through Transdermal Route

Muhammad Rehan Malik, Asif Nawaz and Wajiha Noor

Abstract

Nanoemulsions are colloidal particulate systems that have gained significant attention in pharmaceutical research due to their potential in enhancing drug permeation through transdermal routes. This chapter provides a comprehensive overview of nanoemulsions, including their definition, advantages over other dosage forms, components, methods of preparation, and applications, with a focus on multiple nanoemulsions. The advantages of nanoemulsions include increased absorption rate, reduced absorption variability, improved drug bioavailability, and efficient delivery of both hydrophilic and lipophilic drugs. Various components such as oil phase, surfactants, co-surfactants, aqueous phase, and co-solvents are discussed in detail, along with their roles in nanoemulsion formulation. Methods of nanoemulsion preparation, including high-pressure homogenization, ultrasonic emulsification, high-shear stirring, microfluidization, and membrane emulsification, are elaborated upon, highlighting their respective advantages and limitations. Additionally, applications of nanoemulsions in medicine, catalysis, and theranostics are explored, emphasizing the potential of multiple nanoemulsions in drug delivery systems. Overall, nanoemulsions offer a promising approach for improving drug delivery efficiency and enhancing therapeutic outcomes.

Keywords: nanoemulsion, multiple nanoemulsion, method of preparation, topical delivery cargo, biomedical applications

1. Introduction

1.1 Nanoemulsion

The nanoemulsion technique is one of the most promising strategies for enhancing drug permeation through transdermal route. Submicron emulsions, ultrafine emulsions, and miniemulsions are the term used for nanoemulsion. It may be defined as a system: A submicron sized colloidal particulate system, thermodynamically or kinetically stable isotopically clear dispersions of an oil phase and water phase. The two immiscible liquids are stabilized by an interfacial film of a suitable surfactant and co-surfactant molecules to form a single phase.

A variety of surfactants with different properties that may be ionic or non-ionic have been used with such nanoemulsions. The most commonly used surfactants are sorbitan esters, polysorbates (nonionic surfactants), potassium laurate, sodium lauryl sulphate (anionic surfactants), quaternary ammonium halide (cationic surfactants) etc. The first nanoemulsions were the Oil/water type emulsions, and their droplet size varying from 50 to 1000 nm. The nanoemulsions may be: O/W (oil dispersed in aqueous phase), W/O (water dispersed in oil phase), and multiple emulsions (a type of nanoemulsion that contain both O/W and W/O emulsions in the same system). Both hydrophilic and lipophilic surfactants are used to stabilized such types of systems (i.e., multiple emulsions) simultaneously [1].

1.2 Advantages of nanoemulsions over other dosage forms

1. Increased absorption rate due to vary small droplet size of the nanoemulsion.
2. Reduced absorption variability.
3. In O/W nanoemulsions, protection against oxidation and hydrolysis.
4. Lipophilic drugs delivery after solubilization.
5. Aqueous dosage form for water insoluble drugs.
6. Improved bioavailability of the drugs.
7. Both lipophilic and hydrophilic drugs are incorporated in nanoemulsions.
8. Efficient delivery system for drugs to increase efficiency there by reducing total daily dose and side effects.
9. Nonirritating and non-toxic drug delivery vehicle for skin and mucous membrane.
10. Control release of the drug through nanoemulsion (liquid film) whose thickness, lipophilicity or hydrophilicity may also be controlled.

It not only improves existing emulsion systems, but it also opens up new possibilities for other drugs to be formulated more specifically.

According to the research, nanoemulsion formulations exhibit better cutaneous and transdermal drug transport capabilities with respect to *in-vitro* [2, 3] and *in-vivo* [4, 5] as compared to emulsions [6] and gels [7].

Although nanoemulsion can be used to deliver drugs to patient in no of ways but topical application of nanoemulsion is gaining attention. The mobility of the drug in the vehicle, the release of the drug from the vehicle, and the permeation of the drug into the skin are the three key factors that determine drug transdermal permeation.

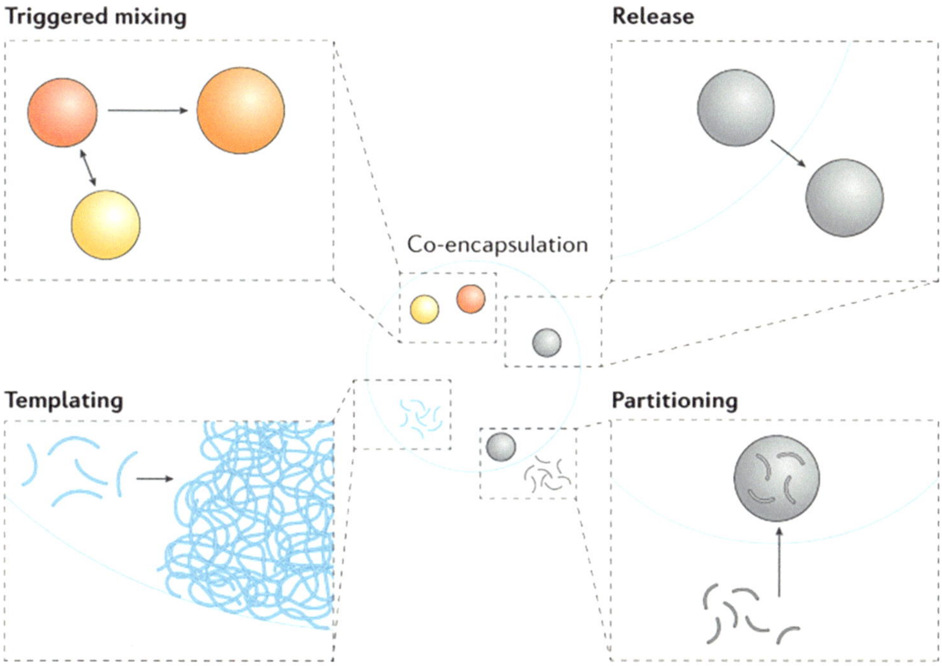

Figure 1.
Design features of multiple nanoemulsions [12].

1.3 Applications of multiple nanoemulsion

Nanoemulsions are used in a variety of fields, such as medicine [8, 9], catalysis [10], and theranostics [11]. Multiple nanoemulsions have distinct advantages for these applications, such as specified phases and internal compartments that can be used to encapsulate multiple cargos at the same time or serve as models for complex nanomaterials (**Figure 1**) Because of their unique structural features and small size, multiple nanoemulsions have significant potential for current nanoemulsion and multiple emulsion applications.

As compared to traditional nanocarriers, [13] optimized an O/W/O multiple nanoemulsion for dermal delivery of antiviral drug aciclovir., which showed efficient skin penetration and better physicochemical stability. Although hydrophilic drugs are more prone to degradation after oral administration, so multiple nanoemulsions can help improve bioavailability in this situation.

As a result, a low-energy method was established by [14] to incorporate hydrophilic anti-cancerous drug 5-FU by formulating multicore W/O/W nanoemulsions. Their formulation showed improved efficacy and absorption efficiency when compared to a simple nanoemulsion oral drug delivery system. This efficient drug delivery and ability to solubilize hydrophilic drugs was achieved by formulating multicore nanoemulsions. The formulated multiple nanoemulsions have the ability to produce a stable drug-delivery system with low surfactant concentrations, neutral pH, and a variety of oils, allowing for long-term shelf stability and drug solubilization.

Both hydrophobic and hydrophilic drugs can be delivered and encapsulated by formulating multiple nanoemulsions, which is not possible with many traditional drug delivery methods [15, 16]. Furthermore, longer pharmacokinetic profile and lower required dosages of both hydrophobic and hydrophilic drugs have been shown due to the durability of the multiple nanoemulsions as compared to conventional drug delivery systems [14, 17].

1.4 Components of nanoemulsion

Drug, oil and aqueous phases, surfactants and co-surfactants are used to prepare nanoemulsion. The physical and chemical properties of these components are taken into account in the formulation, their efficiency and *in-vitro* and *in-vivo* stability.

1.4.1 Oil phase

The choice of other ingredients in nanoemulsions is highly dependent on oil phase. It is critical to choose the right oil for nanoemulsion formulation in order to achieve the desired properties. The oil to be selected for preparation of desired nanoemulsion is highly reliant on the solubility of particular drug. A combination of fixed oil and medium chain triglyceride can be used in order to achieve strong drug loading and

List of oils used in nanoemulsion	
Sr. #	**Oils**
1	Captex 355 (Glyceryl Tricaorylate/Caprate)
2	Captex 200 (Propylene Dicaprylate Dicaprate Glycol)
3	Captex 8000 (Glyceryl Tricaprylate (Tricaprylin)
4	Olive oil
5	Ethyl oleate
6	Isopropyl myristate (Myristic acid isopropyl ester)
7	Soya been oil
8	Glyceryl triacetate
9	Sefsol 218 (Caprylic/Capric Triglyceride)
10	Isopropyl myristate (Tetradecanoic acid)
11	Sesame oil
12	Corn oil
13	Intermediate Chain Triglycerides (Labrafac)
14	Peceol (Glyceryl Oleate)
15	Methyl decanoate
16	Labrafac (medium chain triglyceride)
17	Maisine 35-1 (1-Monolinolein)
18	Capryol 90 (Propylene Glycol Monocaprylate)
19	Capmul MCM (Glycerol monocaprylate)
20	Witepsol
21	Myritol 318

Table 1.
List of oils used as oil phase in nanoemulsion formulation [19].

emulsification. Only those oils should be used that resist auto-oxidation [18]. Due to their high lipophilicity and good solvent ability, medium chain triglycerides (MCT) (resistant to auto-oxidation) preferable over long chain triglyceride. Now adays, semi synthetic medium chain triglyceride are mostly used instead of medium chain triglyceride (**Table 1**) [19].

1.4.2 Surfactants

The type of surfactant used is determined by the type of nanoemulsion being prepared. The selection of particular surfactant is highly dependent on hydrophile–lipophile balance (HLB). Surfactants having low HLB value (<10) are lipophilic in nature and are used to formulate w/o nanoemulsions, while surfactants having high HLB (>10) value are hydrophilic in nature and are used to formulate o/w nanoemulsions (**Figure 2**).

To obtain nanoemulsions, sometimes a mixture of low HLB and high HLB surfactants can be needed. Surfactants are classified into 4 types to formulate a stable naoemulsion [19].

- Ionic surfactants,
- Cationic surfactants,
- Non-ionic surfactants,
- Zwitterionic surfactants (**Table 2**).

1.4.3 Co-surfactants

When surfactant fails to produce a stable formulation, Cosurfactant is usually used to lower the oil–water interfacial tension. Liquid crystalline phases form as the surfactant film becomes too stiff. By entering into the surfactant monolayer and destroying the liquid crystalline phases, co-surfactants increase the fluidity of the interfacial film [19]. Actually, low concentration of co-surfactant is required to formulate nanoemulsion. Short and medium chain alcohol (C3–C8) are often used as co-surfactant. They actually reduce the interfacial tension and increase the fluidity of interface of the nanoemulsion system (**Table 3**) [8].

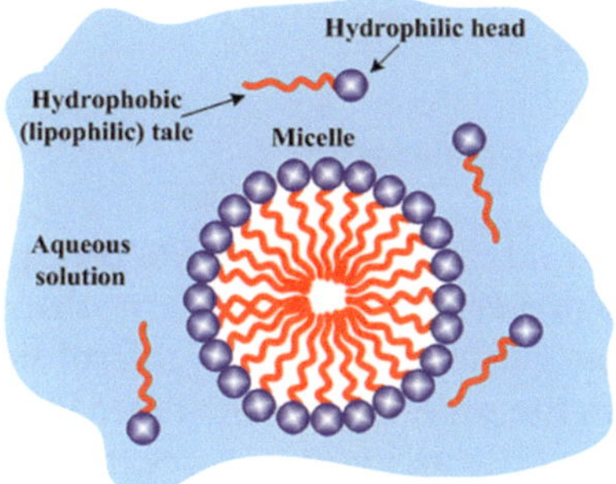

Figure 2.
Surfactant molecular structure [20].

List of surfactants used in nanoemulsions	
Sr. #	**Name of surfactants**
1	Tween 80 (Polyoxyethylene (20) sorbitan monooleate)
2	Tween 20 (Polyoxyethylene (20) sorbitan monolaurate)
3	Span 80 (Sorbitan monooleate)
4	Span 20 (Sorbitan monolaurate)
5	Span 60 (Sorbitan monostearate)
6	Cremophor RH 40 (Polyoxyl 40 hydrogenated castor oil)
7	Poloxamer 407 (Poly (ethylene glycol)-block-poly(propylene glycol)-block-poly(ethylene glycol).
8	Poloxamer 188 ((Poly (ethylene glycol)-block-poly(propylene glycol)-block-poly(ethylene glycol).
9	PEG 400 (Polyethylene glycol 400)
10	PEG 4000 (Polyethylene glycol 4000)
11	PEG 6000 (Polyethylene glycol 6000)
12	Capryol 90
13	Emulphor-620

Table 2.
List of surfactants used to formulate nanoemulsions [19].

1.4.4 Aquous phase

The nature of aquous phase also affect the droplet size and stability of nanoemulsion, like pH, electrolytes and ionic content of aquous phase. Distal water, ringer's solution, simulated gastric and intestinal fluids and phosphate buffer saline will also be used as aquous phase for nanoemulsion preparation. Similarly pH of aquous phase also effect stabilty and phase behavior of nanoemulsion, specially when a drug having pH dependent solubility is used in the formulation [8].

1.4.5 Co-solvents

Sometimes co-solvents are also used to formulate nanoemulsion so that to improve solubilty of either drug or hydrophilic surfactent in the oil phase. They make the formulation more lipophilic by lowering the water's dielectric constant. Organic solvents are mostly used as co-solvents such as glycerol, gasoline, propylene glycol (PG), and polyethylene glycol (PEG) [8].

1.5 Methods of nanoemulsion preparation

High and low energy methods are usually used to prepare nanoemulsion or combination of both. In case of high energy methods, a large distracting force is applied by means of mechanical devices like ultrasonicators, microfuidizers, and high-pressure homogenizers. This result in the production of vary small size droplets. However, Low-energy approaches do not use an external force to produce nanoemulsions; instead, they depend on the system's intrinsic physiological properties. The high energy methods are discussed one by one [8, 21].

List of co-surfactants used in nanoemulsion	
Sr. #	**Name of co-surfactants**
1	Propyl alcohol ($CH_3CH_2CH_2OH$)
2	Ethyl alcohol (CH_3CH_2OH)
3	Transcutol P (Diethylene glycol monoethyl ether)
4	Propylene glycol (Propane-1,2-diol)
5	Glycerin or Glycerol ($CH_2OHCHOHCH_2OH$)
6	Polyglyceryl oleate (poly (oxyethylene) 1,2,3-propanetriyl ester)
7	Propylene glycol laurate (propylene glycol dodecanoate)
8	Apricot kernel oil PEG-6 esters (O-CH_2-CH_2) n where n = 6
9	Propylene glycol monolaurate ($CH_3(CH_2)10COOCH_2CH(OH)CH_2OH$)
10	Medium chain mono- and diglycerides of caprylic acid (2,3-dihydroxypropyl octanoate (for monoglyceride); 1,2,3-propanetriol, di(caprylate) (for diglyceride)
11	Diethylene Glycol Monoethyl ether (2-(2-ethoxyethoxy) ethanol))

Table 3.
List of co-surfactants used in nanoemulsion [19].

1.5.1 High-pressure homogenization

This technique can be used to create nanoemulsions with droplet sizes as small as 1 nm (**Figure 3**). In high-pressure homogenization process, many forces are involved in the creation of nanoemulsion which was achieved by using high pressure homogenizer or piston homogenizer. Cavitation, a lot of turbulence, and hydraulics are a few of them. For the development of nanoemulsion droplets of small size, this process necessitates the use of high energy. Different techniques may be used to improve emulsification. Instead of using a single surfactant, mixture of several surfactants is more applicable so that to reduce surface tension. In order to achieve further small size droplets of the formulation, surfactants may be dissolved in the disperse phase rather than dispersion medium [23]. This method is applicable at both large- and small-scale production of naoemulsion. The method is appealing and effective, but high energy loss as the temperature rises during the manufacturing process (**Figure 4**) [24, 25].

1.5.2 Ultrasonic emulsification

Actually, ultrasonic emulsification is very effective at decreasing the size of droplets. The energy for ultrasonic emulsification is provided using sonotrodes called sonicator probes. It has a piezoelectric quartz crystal that expands and contracts in response to changing electric voltage. A mechanical vibration is produced as the sonicator tip comes in contact with the formulation, resulting in cavitation. The formation and collapse of vapors cavities in formulation is known as cavitation. Therefore, ultrasound can be used directly to formulate nanoemulsion. This method is mostly applicable on small scale production of nanoemulsion, where emulsion droplet sizes as small as 0.2 μm can be achieved [8]. This method is preferred because less concentration of surfactant and less energy is required to formulate nanoemulsion of homogenous nature, compared to other methods [26].

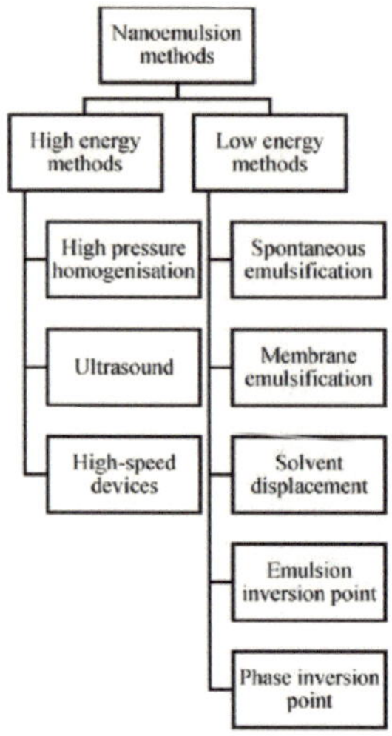

Figure 3.
Methods of nanoemulsion preparation [22].

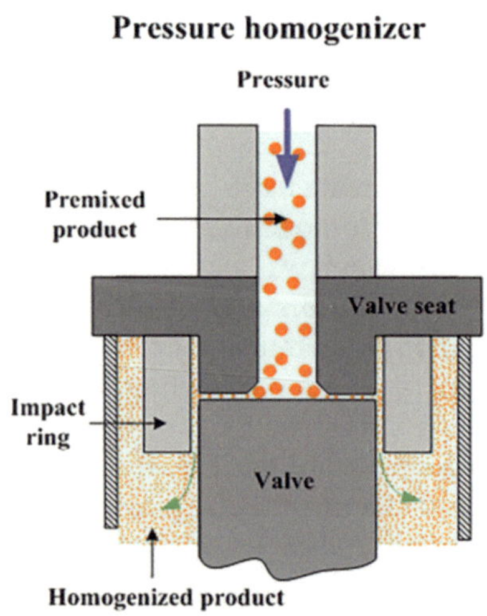

Figure 4.
Pressure homogenizer [24].

1.5.3 High-shear stirring

Nanoemulsions were first made with high-shear mixers and rotor-stator devices (**Figure 5**). High shear stirring is also known as high shear mixer, high shear homogenizer, rotar-stator mixer and high shear reactor. The internal phase droplet size can be reduced dramatically by raising the mixing strength in these machines, but it is difficult to prepare emulsions with average droplet sizes <200–300 nm. Colloid mills are used to achieve increase shear stress during dispersion. The most famous are Silverson flow mixers, which have different configurations for rotors and stators to accomplish more effective emulsification [28].

Ist of all as a result of high rotar speed, a high rotational speed is generated inside crumbling unit and the emulsion molecules are sucked into the rotor-stator assembly.

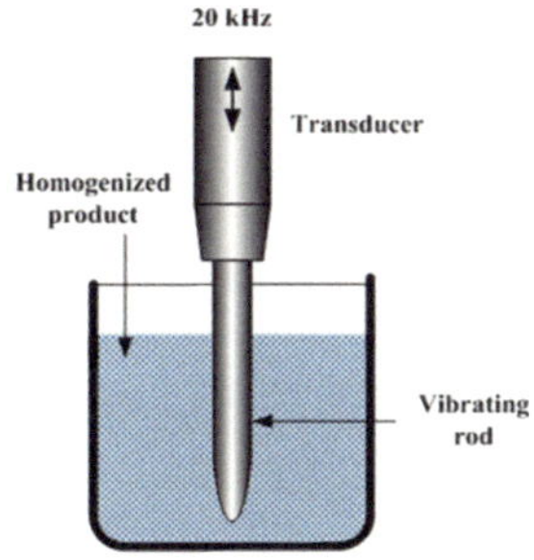

Figure 5.
Ultrasonic homogenizer [27].

Figure 6.
high-shear mixer [30].

After this the emulsion is thrown away to the periphery by centrifugal force, causing extreme turbulance, agitation, and dispersion in the gap between inner rotor and outer stator. Meanwhile, the emulsion flow at high speed over the stator's outer opening and leave the assembly. When the efficiency of high- shear stirring decrease dramatically while using viscous media in emulsion, the portion of internal phase increased and size of globules may exceed to 1 micrometer [29].

1.5.4 Microfluidization

In this process a device known as microfluidizer which employs a high-pressure positive displacement pump (500–20,000 psi), is used to produce nanoemulsion of vary small size (**Figure 6**). This positive displacement pump, pump the fluid into a specially designed chamber (consist of microchannels) at very high speed. After that, the product pushes through microchannels into a collission area, where very small particles in the nm range are formed. In this method, a microemulsion that has already been prepared is inserted into the microfluidizer, resulting in a nanoemulsion with the desired properties is obtained (**Figure 7**) [31].

1.5.5 Membrane emulsification

Membrane emulsification has gotten a lot of attention in the last 10 years, and it has a lot of possible applications. In this process, the internal phase is squeeze out through microchannels or small pores in a membrane having a uniform pore-size distribution into the dispersion medium using a low pressure. The size of the resulting globules is mainly determined by the membrane used, rather than agitation. This

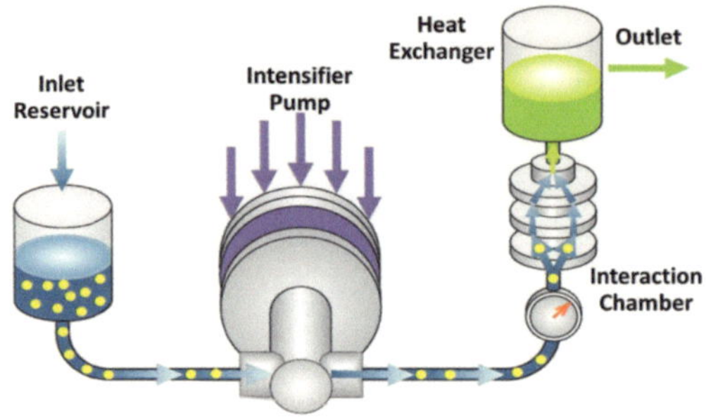

Figure 7.
microfluidization method [31].

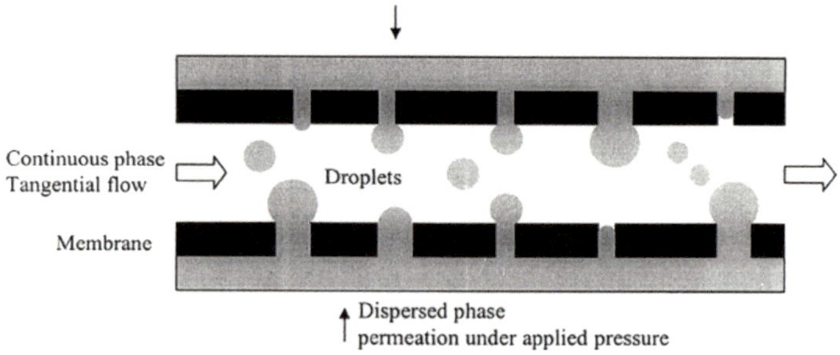

Figure 8.
Schematic diagram of membrane emulsification process [33].

method is highly applicable because of its easiness, use of less surfactant conc and lower energy requirement, and very small/uniform particle-size distributions. This technique is used to formulate both o/w and w/o emulsions [32].

2. Conclusion

Nanoemulsions represent a versatile and promising platform for drug delivery, offering numerous advantages over traditional dosage forms (**Figure 8**). The ability to encapsulate both hydrophilic and lipophilic drugs, along with their enhanced bioavailability and controlled release properties, makes nanoemulsions an attractive option for pharmaceutical formulations. Furthermore, the development of multiple nanoemulsions opens up new possibilities for complex drug delivery systems with specified phases and internal compartments, allowing for the encapsulation of multiple cargos simultaneously. Despite the challenges associated with nanoemulsion formulation and preparation, ongoing research efforts continue to explore novel methods and applications to further enhance their effectiveness and utility in various fields. Overall, nanoemulsions hold great promise for advancing drug delivery technologies and improving patient outcomes.

Acknowledgements

All praises and glory be to Allah Almighty, the Lord of the world, the Master of the Day After, Who gave me the courage and strength to carry out this work and made everything possible for me to complete my M.Phil degree in Pharmaceutics.

Peace and blessing be upon the Holy Prophet Muhammad ﷺ whose way of life is continue guidance for mankind and source of knowledge for humanity as a whole.

I would like to express my special appreciation and thanks to my research supervisor Dr. Asif Nawaz for his assistance, guidance, help and dedicated involvement in every step of my research.

I am very grateful and say thanks to my teacher Dr. Akhlaq, Dr. Barkat for their advice, support and encouragement during my research work.

Author details

Muhammad Rehan Malik[1,2]*, Asif Nawaz[1] and Wajiha Noor[3]

1 Faculty of Pharmacy, Gomal Center of Pharmaceutical Sciences, Gomal University, Dera Ismail Khan, KP, Pakistan

2 Department of Pharmacy, The University of Chenab, Gujarat, Pakistan

3 University of Lahore, Gujrat Campus, Pakistan

*Address all correspondence to: malicrehan@gmail.com

References

[1] Gurpreet K, Singh SJ. Review of nanoemulsion formulation and characterization techniques. Indian Journal of Pharmaceutical Sciences. 2018;**80**(5):781-789

[2] Delgado-Charro MB et al. Delivery of a hydrophilic solute through the skin from novel microemulsion systems. European Journal of Pharmaceutics and Biopharmaceutics. 1997;**43**(1):37-42

[3] Kreilgaard EJP, Jaroszewski JW. NMR characterisation and transdermal drug delivery potential of microemulsion systems. Journal of Controlled Release. 2000;**69**(3):421-433

[4] Lee PJ, Langer R, Shastri VP. Novel microemulsion enhancer formulation for simultaneous transdermal delivery of hydrophilic and hydrophobic drugs. Pharmaceutical Research. 2003;**20**(2):264-269

[5] Kreilgaard EJ. Dermal pharmacokinetics of microemulsion formulations determined by in vivo microdialysis. Journal of Pharmaceutical Research. 2001;**18**(3):367-373

[6] Kreilgaard EJ et al. Influence of a microemulsion vehicle on cutaneous bioequivalence of a lipophilic model drug assessed by microdialysis and pharmacodynamics. Pharmaceutical Research. 2001;**18**(5):593-599

[7] Kriwet K, Müller-Goymann CC. Diclofenac release from phospholipid drug systems and permeation through excised human stratum corneum. 1995;**125**(2):231-242

[8] Jaiswal M, Dudhe R, Sharma PJB. Nanoemulsion: an advanced mode of drug delivery system. 2015;**5**(2):123-127

[9] Hamidi M, Azadi A, Rafiei PJ. Hydrogel nanoparticles in drug delivery. 2008;**60**(15):1638-1649

[10] Gawande MB et al. Core–shell nanoparticles: synthesis and applications in catalysis and electrocatalysis. Chemical Society Reviews. 2015;**44**(21):7540-7590

[11] Patel MJ, Pollock JA, Sravan Kumar K, Janjic JM. Two-color fluorescent (near-infrared and visible) triphasic perfluorocarbon nanoemulsions. Journal of Biomedical Optics. 2013;**18**(10):101312

[12] Sheth T et al. Multiple nanoemulsions. Nature Reviews Materials. 2020;**5**(3):214-228

[13] Schwarz JC et al. Optimisation of multiple W/O/W nanoemulsions for dermal delivery of aciclovir. 2012;**435**(1):69-75

[14] Shakeel F et al. Double w/o/w nanoemulsion of 5-fluorouracil for self-nanoemulsifying drug delivery system. 2014;**200**:183-190

[15] Wu S-H, Hung Y, Mou C-Y. Compartmentalized hollow silica nanospheres templated from nanoemulsions. 2013;**25**(3):352-364

[16] Ding S et al. A new method for the formulation of double nanoemulsions. 2017;**13**(8):1660-1669

[17] Pangeni R et al. Multiple nanoemulsion system for an oral combinational delivery of oxaliplatin and 5-fluorouracil: preparation and in vivo evaluation. 2016;**11**:6379

[18] Jumaa M, Mueller B. Formulating and stability of benzodiazepines in a new lipid emulsion formulation. 2002;**57**(11):740-743

[19] Sarker A et al. Nanoemulsion: An excellent mode for delivery of poorly soluble drug through different routes. 2015;**7**(12):966-976

[20] Amirpour M et al. Experimental investigation of wettability alteration on residual oil saturation using nonionic surfactants: Capillary pressure measurement. 2015;**1**(4):289-299

[21] Shakeel F et al. Stability evaluation of celecoxib nanoemulsion containing Tween. 2008;**80**(32):4-9

[22] Banasaz S et al. Encapsulation of lipid-soluble bioactives by nanoemulsions. 2020;**25**(17):3966

[23] Lovelyn C, Attama AA. Current state of nanoemulsions in drug delivery. 2011;**2**(05):626

[24] Debnath S, Satayanarayana KV, Kumar GV. Nanoemulsion—a method to improve the solubility of lipophilic drugs. Pharmanest. 2011;**2**(2-3):72-83

[25] Nor Bainun I, Alias NH, Syed-Hassan SSA. Nanoemulsion: Formation, Characterization, Properties and Applications: A Review in Advanced Materials Research. Trans Tech Publication; 2015

[26] Cucheval A, Chow R. A study on the emulsification of oil by power ultrasound. 2008;**15**(5):916-920

[27] Babajide O et al. Low-cost feedstock conversion to biodiesel via ultrasound technology. 2010;**3**(10):1691-1703

[28] Koroleva MY, Yurtov EV. Nanoemulsions: The properties, methods of preparation and promising applications. 2012;**81**(1):21

[29] Shah M, Imran M, Ullah SJL. Nanoemulsions. 2017:111-137

[30] Zhang J et al. High shear mixers: A review of typical applications and studies on power draw, flow pattern, energy dissipation and transfer properties. 2012;**57**:25-41

[31] Talegaonkar S, Negi LM. Nanoemulsion in drug targeting. In: Targeted Drug Delivery: Concepts and Design. Springer; 2015. pp. 433-459

[32] Joscelyne SM, Trägårdh G. Membrane emulsification—a literature review. 2000;**169**(1):107-117

[33] Charcosset C et al. The membrane emulsification process—a review. 2004;**79**(3):209-218